SpringerBriefs in Applied Sciences and Technology

SpringerBriefs present concise summaries of cutting-edge research and practical applications across a wide spectrum of fields. Featuring compact volumes of 50 to 125 pages, the series covers a range of content from professional to academic.

Typical publications can be:

- A timely report of state-of-the art methods
- An introduction to or a manual for the application of mathematical or computer techniques
- A bridge between new research results, as published in journal articles
- A snapshot of a hot or emerging topic
- An in-depth case study
- A presentation of core concepts that students must understand in order to make independent contributions

SpringerBriefs are characterized by fast, global electronic dissemination, standard publishing contracts, standardized manuscript preparation and formatting guidelines, and expedited production schedules.

On the one hand, **SpringerBriefs in Applied Sciences and Technology** are devoted to the publication of fundamentals and applications within the different classical engineering disciplines as well as in interdisciplinary fields that recently emerged between these areas. On the other hand, as the boundary separating fundamental research and applied technology is more and more dissolving, this series is particularly open to trans-disciplinary topics between fundamental science and engineering.

Indexed by EI-Compendex, SCOPUS and Springerlink.

Alaa M. Rashad

Metakaolin Effect on Geopolymers' Properties

 Springer

Alaa M. Rashad
Housing and Building National Research
Center (HBRC)
Building Materials and Quality Control
Institute
Cairo, Egypt

ISSN 2191-530X   ISSN 2191-5318 (electronic)
SpringerBriefs in Applied Sciences and Technology
ISBN 978-3-031-45150-8   ISBN 978-3-031-45151-5 (eBook)
https://doi.org/10.1007/978-3-031-45151-5

This Springer imprint is published by the registered company Springer Nature Switzerland AG
The registered company address is: Gewerbestrasse 11, 6330 Cham, Switzerland

Paper in this product is recyclable.

Preface

The World Health Organization (WHO) reported that 9 out of 10 people inhale air with pollution levels that exceed the acceptable limit outlined by the WHO. This contributes to a global burden of approximately 7 million deaths per year caused by inhalation of particle dust from polluted air [1]. Unfortunately, the quality of air is negatively impacted by the cement industry due to the emission of particulate matter, SO_2, CO_2 and NOx. This industry is one of the major sources of air pollution, contributing significantly to various health issues, including cardiopulmonary sickness, myocardial infection, respiratory infections and ischaemic stroke [2]. The cement industry utilizes a significant amount of energy compared to other industrial sectors, accounting for approximately 15% of the total energy [3]. The cement industry also relies heavily on natural raw materials, but efforts have been made to mitigate these negative impacts by utilizing alternative cementitious materials. One such effective solution is to completely replace cement with more environmentally friendly binders such as geopolymers. Geopolymers can be produced by activating a precursor with a high aluminosilicate content using an alkali activator and have the potential to eliminate the adverse effects associated with cement production.

In 1930, Kuhl conducted a study on the setting behaviour of powdered slag activated with a solution of caustic potash, which marked the first use of alkali cement. Later, in 1937, Chassevent used caustic potash and soda to evacuate the reactivity of powdered slag [4]. In 1940, Purdon [5] carried out the first inclusive laboratory investigation on cements made without clinker utilizing slag and either caustic soda or caustic alkalis produced by mixing a base and an alkaline. In 1957, Glukhovsky developed binders using low-basic calcium aluminosilicates activated with an alkali metal solution or calcium-free materials [6]. Davidovits developed binders in 1981 by combining alkalis with a burnt mixture of kaolinite, dolomite and limestone. These binders were termed "geopolymers" due to their polymeric structure. Moreover, Davidovits employed several trademarks, including Pyrament, Geopolyccm and Geopolymite, for the binders [4]. Malek and his team [7] recognized alkali-activated cement materials as the matrix that develops during the solidification of specific radioactive wastes in 1986. In 1989, Roy and Langton [8] discovered similarities between such materials and ancient concretes. Krivenko's [9] research in

1994 demonstrated that when the concentration of alkalis is adequate, alkalis, alkali metal salts, silicates, aluminates and aluminosilicates react in alkaline aqueous environments. This reaction occurs with clay minerals, aluminosilicate glasses from natural and manmade sources that lack calcium, as well as with natural calcium-binding systems. As a result, water-resistant hardened alkali or alkali alkaline earth hydroaluminosilicates are formed. These hydroaluminosilicates are comparable to natural zeolites and micas. Since then, there has been increasing interest in manufacturing alkali-activated materials or geopolymers due to the rising demand for innovative building materials that have low greenhouse gas emissions during production. Geopolymers or alkali-activated materials are commonly produced by using an alkaline activator to activate aluminosilicate materials such as metakaolin, fly ash or slag.

Metakaolin is prepared through regulated thermal processing of kaolinite (kaolin), with specific temperatures in the range of 600–850 °C and durations in the range of 1–12 h [10]. Metakaolin has a high pozzolanic index, which enables it to be used as a part of cement in Portland cement systems to enhance strength development, improve durability and decrease permeability. Because metakaolin is a rich material in alumina and silica, it can be activated by alkalis to produce cementing material [11]. The reaction of metakaolin with alkalis is slow at room temperature. This defect opens the door to the use of metakaolin as an additive or a part of the main precursor(s). The incorporation of metakaolin as a part of precursor(s) may have a positive effect on some properties or a negative effect on others.

In this book, the results of the effect of metakaolin on different types of geopolymers were collected and analysed. Chapter 1 concentrated on highlighting the significance of adopting geopolymers as a replacement for Portland cement, elaborating on the importance of metakaolin utilization in both Portland cement and geopolymers and examining the properties of metakaolin. Chapter 2 concentrated on the effect of metakaolin on the fresh properties of different types of geopolymers, such as workability and setting time. Chapter 3 concentrated on the effect of metakaolin on some physical properties of different types of geopolymers, such as density, porosity, water absorption and shrinkage. Chapter 4 concentrated on the effect of metakaolin on the mechanical properties of different types of geopolymers, such as compressive strength, flexural strength, splitting tensile strength and elastic modulus. Chapter 5 concentrated on the effect of metakaolin on the durability of different types of geopolymers. Chapter 6 provided a comprehensive overview of the outcomes discussed in the preceding chapters, along with suggestions for future research. Finally, Chap. 7's primary objective was to provide an overview of the general observations or conclusions regarding the incorporation of metakaolin into different types of geopolymers.

Cairo, Egypt   Alaa M. Rashad

References

1. E. Adeyanju, C.A. Okeke, Exposure effect to cement dust pollution: a mini review, SN Appl. Sci. **1**(12), 1572 (2019)
2. M. Kampa, E. Castanas, Human health effects of air pollution. Environ. Pollut. **151**(2), 362–367 (2008)
3. A. Mokhtar, M. Nasooti, A decision support tool for cement industry to select energy efficiency measures. Energy Strategy Rev. **28** (2020) 100458
4. C. Shi, D. Roy, P. Krivenko, Alkali-activated cements and concretes, CRC Press (2003)
5. A. Purdon, The action of alkalis on blast-furnace slag. J. Soc. Chem. Indus. **59**(9), 191–202 (1940)
6. V. Glukhovsky, Soil silicates (Gruntosilikaty), Kiev, USSR Budivelnik Publ (1959)
7. R.I. Malek, D. Roy, C. Langton, Slag cement-low level radioactive waste forms at Savannah River Plant, Am. Ceram. Soc. Bull. (United States) **65**(12) (1986)
8. D.M. Roy, C. Langton, Studies of Ancient Concrete as Analogs of Cementitious Sealing Materials for a Repository in Tuff, Los Alamos National Lab.(LANL) (Los Alamos, NM (United States), 1989)
9. P. Krivenko, Influence of physico-chemical aspects of early history of a slag alkaline cement stone on stability of its properties, in *1st International Conference on Reinforced Concrete Materials in Hot Climates* (1994)
10. A.M. Rashad, Metakaolin as cementitious material: history, scours, production and composition–a comprehensive overview. Construct. Build. Mater. **41**, 303–318 (2013)
11. A.M. Rashad, Alkali-activated metakaolin: a short guide for civil engineer–an overview. Construct. Build. Mater. **41**, 751–765 (2013)

Contents

Chapter 1
Introduction

Although Portland cement (PC) is widely used in construction and infrastructure, the PC industry consumes a tremendous amount of natural resources and energy. To partially address these flaws, blended cement was employed [1–6]. However, the flaws resulting from the cement industry may be completely resolved if other environmentally friendly binders could completely replace cement. Geopolymers are the name of these environmentally beneficial binders. Geopolymer is a binder resulting from the interaction between aluminosilicate and activator. Employing geopolymers rather than PC can reduce energy consumption and greenhouse gas emissions by ~43% and ~73%, respectively [7]. Since the discovery of geopolymers, ameliorations have been made to achieve improved properties. Blending the primary precursor with MK is one option to achieve greater qualities.

A dehydroxylated version of the clay mineral kaolinite is known as metakaolin (MK) [8]. Kaolin or kaolinite is a natural mineral that occupies a large part of the Earth's crust (Fig. 1.1). North America, South America, several European nations, Australia, China, Southeast Asia and the Indian Subcontinent all have kaolin reserves [9]. In addition, kaolin reserves are abundant in Egypt in many locations, such as Kalabsh near Aswan, the Red Sea coast and Sinai [10, 11]. According to the United States Geological Survey, worldwide kaolin production was estimated to be approximately 45 million tonnes [12]. Kaolin is a clay mineral that offers raw material plasticity. Structurally, it consists of silica tetrahedral sheets and alumina octahedral sheets with a theoretical composition of ~39.5% alumina, ~46.54% silica and ~13.96% H_2O. Figure 1.2 displays the configuration of atoms in the kaolinite group [13]. The majority of the kaolin deposits consist of kaolinite crystals that take on a pseudohexagonal shape, with both plate-like and larger book-like structures, as well as vermicular stacks (Fig. 1.3) [14]. Kaolin is incredibly beneficial. Due to its softness, small particle size, white colour and chemical inertness, it is suitable for a variety of applications such as the production of porcelain, ceramic, cement, bricks, paper, rubber, paint, plastic, aluminium and pigments. It can also be used for cosmetics, toothpaste and as an additive for food. Currently, a specially prepared spray from

A. M. Rashad, *Metakaolin Effect on Geopolymers' Properties*,
SpringerBriefs in Applied Sciences and Technology,
https://doi.org/10.1007/978-3-031-45151-5_1

kaolin is applied to vegetables and fruits whilst they are still in the developing process to ward against sunburn and insects [15]. Under the use of proper thermal activation (calcination), kaolin can be transformed into MK. The calcination temperature could be in the range of 600–850 °C for 1–12 h. Calcination can produce amorphous material. MK ($Al_2O_3.2SiO_2$) was used for the first time in 1962 in Jupia Dam concrete [16]. Since the middle of the 1990s, it has been accessible through commerce. The colour of MK is usually white, greenish, yellow or colourless (Fig. 1.4) making it very appealing for colour coordination and other architectural uses. The morphology of MK has a predominantly irregular spiny plate-like form (Fig. 1.5). The XRD pattern of the source of MK before calcination (i.e., kaolin) shows the crystalline phases of kaolinite and quartz (Fig. 1.6a). As shown in Fig. 1.6b, after calcination, the XRD pattern of MK presents the coexistence of crystalline and amorphous phases, whilst the phases of kaolinite have disappeared. The amorphous phase is confirmed by the presence of cumber at 20–30° 2θ, whilst the crystalline phases of hematite, quartz and illite can be detected (Fig. 1.6b). The range of MK specific gravity is 2.3 [17], 2.34 [18], 2.35 [19, 20], 2.4 [21–23], 2.5 [24, 25] or 2.6 [26, 27]. MK contains high amounts of alumina and silica in addition to small amounts of other oxides such as CaO, Fe_2O_3, MgO, TiO_2, Na_2O and K_2O as presented in Table 1.1. The particle size of MK particles is smaller than cement but larger than silica fume (SF). Due to its impressive properties, it can be used in ceramic production and as a cement replacement in concrete and mortar. The introduction of MK into PC systems has plateful benefits such as improved workability [28], improved mechanical strength [29], enhanced durability, fire resistance and abrasion resistance [30–33]. It can also be used to produce thermal insulation material [34]. Recently, MK has been used as a precursor material for geopolymer production [13, 35] due to its purity, reactivity, low cost [36], availability and consistent chemical composition [37, 38] or as an additive to other types of geopolymers [39–41]. Despite the literature has scattered and planful studies focused on the effect of MK on the properties of different geopolymer types, so far, there is no document summarizing and analysing these scattered and planned studies. Thus, the target of this document is to briefly analyse earlier studies on the influence of MK on the properties of different types of geopolymers.

Fig. 1.1 Kaolin view [13].
Reprinted with permission
from Elsevier publisher

Fig. 1.2 Kaolinite
schematic structural [13].
Reprinted with permission
from Elsevier publisher

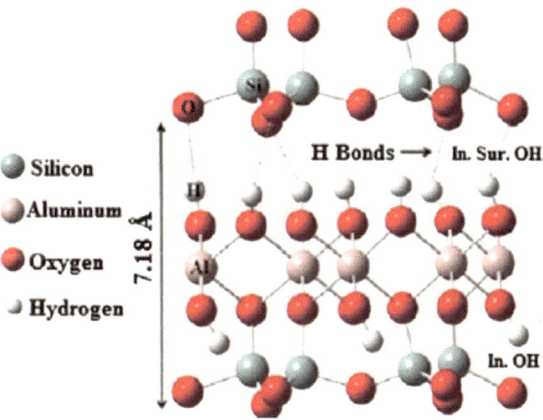

Fig. 1.3 SEM image shows
kaolin stacks and plates [14].
Reprinted with permission
from Elsevier publisher

Fig. 1.4 MK view [42].
Reprinted with permission
from Springer publisher

Fig. 1.5 MK morphology
[38]. Reprinted with
permission from Elsevier
publisher

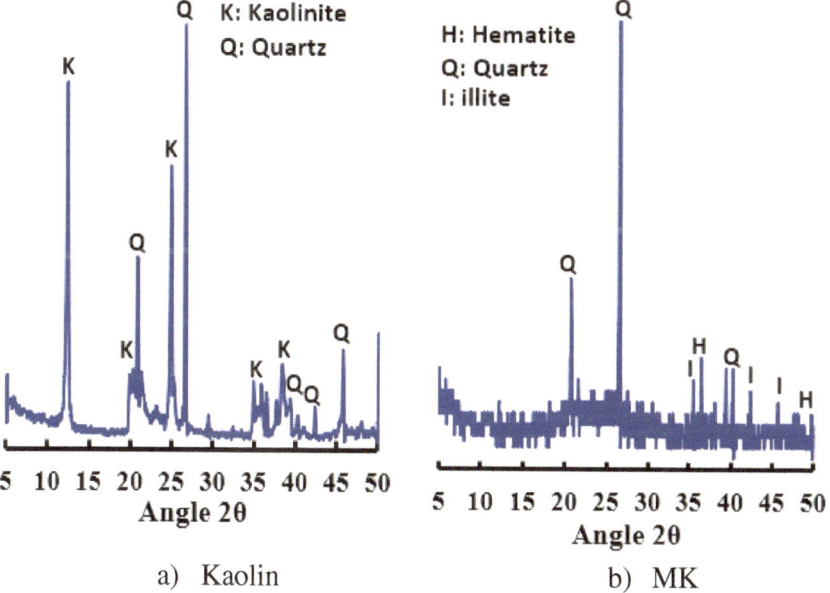

a)   Kaolin   b)   MK

Fig. 1.6 XRD patterns for source kaolin **a** and MK **b** [38]. Reprinted with permission from Elsevier
publisher

Table 1.1 Main chemical composition of MK

References	Oxide (%)											
	SiO_2	Al_2O_3	CaO	Fe_2O_3	MgO	SO_3	P_2O_5	MnO	TiO_2	Na_2O	K_2O	L.O.I
Yurt and Bekar [43]	64.86	20.56	1.24	0.31	0.22	1.01	–	–	0.37	0.11	0.5	10.82
Khalil et al. [44]	61.24	20.89	0.16	1.06	0.22	–	–	–	1.61	0.71	–	13.12
Burciaga-Díaz et al. [45]	51.05	45.26	0.1	0.34	–	–	–	–	1.76	0.06	0.15	–
Chen and Ye [46]	55.31	43.55	–	0.44	–	2.49	–	–	0.7	–	–	1.15
Rashad et al. [47]	58.51	35.55	1.25	1.14	0.18	0.07	–	–	–	0.24	0.06	3
Shi et al. [48]	53.92	41.63	0.68	1.33	0.41	–	–	–	–	0.49	0.34	–
Alanazi et al. [49]	55.01	40.94	0.14	0.55	0.34	–	–	–	0.55	0.09	0.6	1.54
Mahmoodi et al. [21]	55.47	38.07	0.03	–	–	–	–	–	1.47	0.02	0.27	–
Gómez-Casero et al. [50]	54	43	0.1	0.48	0.1	–	0.05	<0.01	0.24	0.01	0.5	0.44
Duan et al. [51]	53.32	42.09	0.09	2.33	0.21	–	–	0.02	0.63	0.49	0.64	0.08
Fu et al. [52]	52.75	44.81	0.02	0.37	0.14	0.02	0.07	–	1.28	0.27	0.14	–
Burciaga-Díaz et al. [53]	51.05	45.26	0.1	0.34	–	–	–	–	1.76	0.06	0.15	–
Rashad and Ouda [54]	52.6	37	0.34	0.42	0.05	0.03	0.32	–	1.61	0.1	0.06	0.32
Buchwald et al. [55]	52.1	43	–	0.7	0.3	–	–	–	–	0.12	2.5	1
Rashad [56]	52.6	37	0.34	0.42	0.05	–	0.55	0.03	1.61	0.1	0.06	6.03
Peng et al. [57]	50.07	48.23	0.24	–	0.22	0.02	–	–	–	0.06	–	–
Pascual et al. [58]	52.37	44.85	0.02	0.42	0.02	–	–	–	–	0.11	0.11	0.45
Zhang et al. [59]	51.35	44.24	0.13	0.98	0.48	–	0.45	0.01	0.9	0.16	0.08	0.72
Rovnaník et al. [60]	58.7	38.5	0.2	0.72	0.38	0.02	–	–	0.49	0.04	0.85	1.67
Gharieb et al. [26]	62.22	31.6	0.26	1.47	0.16	0.15	0.1	–	2.6	0.13	0.12	1.16
Chen and Ye [61]	55.31	43.55	–	0.44	–	–	–	–	0.7	–	–	1.15

(continued)

Table 1.1 (continued)

References	Oxide (%)											
	SiO_2	Al_2O_3	CaO	Fe_2O_3	MgO	SO_3	P_2O_5	MnO	TiO_2	Na_2O	K_2O	L.O.I
Rashad [35]	58.52	35.54	1.24	1.15	0.19	0.06	–	–	–	0.25	0.05	2.74
Burciaga-Díaz et al. [62]	51.05	45.26	0.1	0.34	–	–	–	–	1.76	0.06	0.15	–
Morsy et al. [63, 64]	58.52	35.54	1.24	1.15	0.19	0.06	–	–	–	0.25	0.05	2.74
Rashad and Sadek [4, 65]	47	37	0.2	0.2	0.02	–	–	–	1.3	0.15	0.04	13.4

References

1. A.M. Rashad, A brief on high-volume class F fly ash as cement replacement–a guide for civil engineer. Int. J. Sustain. Built Environ. **4**(2), 278–306 (2015)
2. A.M. Rashad, Potential use of silica fume coupled with slag in HVFA concrete exposed to elevated temperatures. J. Mater. Civ. Eng. **27**(11), 04015019 (2015)
3. A.M. Rashad, An investigation of high-volume fly ash concrete blended with slag subjected to elevated temperatures. J. Clean. Prod. **93**, 47–55 (2015)
4. A.M. Rashad, D.M. Sadek, An investigation on Portland cement replaced by high-volume GGBS pastes modified with micro-sized metakaolin subjected to elevated temperatures. Int. J. Sustain. Built Environ. **6**(1), 91–101 (2017)
5. A.M. Rashad, H.A. El-Nouhy, S.R. Zeedan, An investigation on HVS paste modified with nano-SiO_2 imperiled to elevated temperatures. Arab. J. Sci. Eng. **43**(10), 5165–5177 (2018)
6. A.M. Rashad, An overview on rheology, mechanical properties and durability of high-volume slag used as a cement replacement in paste, mortar and concrete. Constr. Build. Mater. **187**, 89–117 (2018)
7. A.M. Rashad, Additives to increase carbonation resistance of slag activated with sodium sulfate. ACI Mater. J. **119**(2) (2022)
8. A.M. Rashad, Performance of autoclaved alkali-activated metakaolin pastes blended with micro-size particles derivative from dehydroxylation of kaolinite. Constr. Build. Mater. **248**, 118671 (2020)
9. G. Sposito, in *The Chemistry of Soils* (Oxford University Press, 2008)
10. N.A. Abdel-Khalek, The Egyptian kaolin: an outlook in the view of the new climate of investment. Appl. Clay Sci. **15**(3–4), 325–336 (1999)
11. M. Morsy, A.M. Rashad, H. Shoukry, M. Mokhtar, Potential use of limestone in metakaolin-based geopolymer activated with H_3PO_4 for thermal insulation. Constr. Build. Mater. **229**, 117088 (2019)
12. USGS, US-Geological Survey, Mineral commodity-summaries 2022 (2022)
13. A.M. Rashad, Alkali-activated metakaolin: a short guide for civil engineer–an overview. Constr. Build. Mater. **41**, 751–765 (2013)
14. H.H. Murray, Traditional and new applications for kaolin, smectite, and palygorskite: a general overview. Appl. Clay Sci. **17**(5–6), 207–221 (2000)
15. A.M. Rashad, Metakaolin as cementitious material: history, scours, production and composition–a comprehensive overview. Constr. Build. Mater. **41**, 303–318 (2013)
16. M.N. Saad, W.P. De Andrade, V.A. Paulon, Properties of mass concrete containing an active Pozzolan made from clay. Concr. Int. **4**(7), 59–65 (1982)
17. H. Guo, B. Zhang, L. Deng, P. Yuan, M. Li, Q. Wang, Preparation of high-performance silico-aluminophosphate geopolymers using fly ash and metakaolin as raw materials. Appl. Clay Sci. **204**, 106019 (2021)
18. A. Rashad, Chloride ion permeability of plain and blended cement concretes. Inst. Constr. Arch. Slovak Acad. Sci. Build. Res. J. BRJ **59**(1–2), 39–52 (2011)
19. A. Rashad, M. Morsy, H. El-Nouhy, Effect of elevated temperature on physico-mechanical properties of metakaolin blended cement mortar. Struct. Eng. Mech. Int'l J. **31**(1), 1–10 (2009)
20. A. Rashad, Effect of incorporating blended cement on fresh and hardened properties of concrete. Silicates Indus (7–8), 207–215 (2009)
21. O. Mahmoodi, H. Siad, M. Lachemi, S. Dadsetan, M. Sahmaran, Development of ceramic tile waste geopolymer binders based on pre-targeted chemical ratios and ambient curing. Constr. Build. Mater. **258**, 120297 (2020)
22. B. El Moustapha, S. Bonnet, A. Khelidj, N. Leklou, D. Froelich, I.A. Babah, C. Charbuillet, A. Khalifa, Compensation of the negative effects of micro-encapsulated phase change materials by incorporating metakaolin in geopolymers based on blast furnace slag. Constr. Build. Mater. **314**, 125556 (2022)

23. B. El Moustapha, S. Bonnet, A. Khelidj, N. Maranzana, D. Froelich, A. Khalifa, I.A. Babah, Effects of microencapsulated phase change materials on chloride ion transport properties of geopolymers incorporating slag and, metakaolin, and cement-based mortars. J. Build. Eng., 106887 (2023)

24. V.V. Praveen Kumar, N. Prasad, S. Dey, Influence of metakaolin on strength and durability characteristics of ground granulated blast furnace slag based geopolymer concrete. Struct. Concr. **21**(3), 1040–1050 (2020)

25. A.M. Rashad, Possibility of producing thermal insulation materials from cementitious materials without foaming agent or lightweight aggregate. Environ. Sci. Pollut. Res. **29**(3), 3784–3793 (2022)

26. M. Gharieb, Y.A. Mosleh, A.M. Rashad, Properties and corrosion behaviour of applicable binary and ternary geopolymer blends. Int. J. Sustain. Eng. **14**(5), 1068–1080 (2021)

27. N. Bheel, P. Awoyera, T. Tafsirojjaman, N.H. Sor, Synergic effect of metakaolin and groundnut shell ash on the behavior of fly ash-based self-compacting geopolymer concrete. Constr. Build. Mater. **311**, 125327 (2021)

28. A.M. Rashad, Metakaolin: fresh properties and optimum content for mechanical strength in traditional cementitious materials-a comprehensive overview. Rev. Adv. Mater. Sci. **40**(1), 15–44 (2015)

29. A.M. Rashad, H.E.D. Seleem, K.M. Yousri, Compressive strength of concrete mixtures with binary and ternary cement blends. Build. Res. J. **57**(2), 107–130 (2009)

30. A.M. Rashad, A synopsis about the effect of metakaolin on the durability of Portland cement-an overview (2015)

31. H.E.-D.H. Seleem, A.M. Rashad, B.A. El-Sabbagh, Durability and strength evaluation of high-performance concrete in marine structures. Constr. Build. Mater. **24**(6), 878–884 (2010)

32. H.E.D.H. Seleem, A.M. Rashad, T. Elsokary, Effect of elevated temperature on physico-mechanical properties of blended cement concrete. Constr. Build. Mater. **25**(2), 1009–1017 (2011)

33. A.M. Rashad, A preliminary study on the effect of fine aggregate replacement with metakaolin on strength and abrasion resistance of concrete. Constr. Build. Mater. **44**, 487–495 (2013)

34. A.M. Rashad, Possibility of using metakaolin as thermal insulation material. Int. J. Thermophys. **38**, 1–11 (2017)

35. A.M. Rashad, Insulating and fire-resistant behaviour of metakaolin and fly ash geopolymer mortars. Proc. Inst. Civ. Eng. Constr. Mater **172**(1), 37–44 (2019)

36. L. Li, J. Xie, B. Zhang, Y. Feng, J. Yang, A state-of-the-art review on the setting behaviours of ground granulated blast furnace slag-and metakaolin-based alkali-activated materials. Constr. Build. Mater. **368**, 130389 (2023)

37. A.M. Rashad, A.S. Ouda, D.M. Sadek, Behavior of alkali-activated metakaolin pastes blended with quartz powder exposed to seawater attack. J. Mater. Civ. Eng. **30**(8), 04018159 (2018)

38. A.M. Rashad, M. Gharieb, H. Shoukry, M. Mokhtar, Valorization of sugar beet waste as a foaming agent for metakaolin geopolymer activated with phosphoric acid. Constr. Build. Mater. **344**, 128240 (2022)

39. A.M. Rashad, A comprehensive overview about the influence of different additives on the properties of alkali-activated slag–a guide for civil engineer. Constr. Build. Mater. **47**, 29–55 (2013)

40. A.M. Rashad, A comprehensive overview about the influence of different admixtures and additives on the properties of alkali-activated fly ash. Mater. Des. **53**, 1005–1025 (2014)

41. A.M. Rashad, M.H. Khalil, M. El-Nashar, Insulation efficiency of alkali-activated lightweight mortars containing different ratios of binder/expanded perlite fine aggregate. Inno. Infrastruct. Solutions **6**(3), 156 (2021)

42. A.M. Rashad, G.M. Essa, W. Morsi, Traditional cementitious materials for thermal insulation. Arab. J. Sci. Eng., **47**(10), 12931–12943 (2022)

43. Ü. Yurt, F. Bekar, Comparative study of hazelnut-shell biomass ash and metakaolin to improve the performance of alkali-activated concrete: a sustainable greener alternative. Constr. Build. Mater. **320**, 126230 (2022)

44. M.G. Khalil, F. Elgabbas, M.S. El-Feky, H. El-Shafie, Performance of geopolymer mortar cured under ambient temperature. Constr. Build. Mater. **242**, 118090 (2020)
45. O. Burciaga-Díaz, L.Y. Gómez-Zamorano, J.I. Escalante-García, Influence of the long term curing temperature on the hydration of alkaline binders of blast furnace slag-metakaolin. Constr. Build. Mater. **113**, 917–926 (2016)
46. Z. Chen, H. Ye, Influence of metakaolin and limestone on chloride binding of slag activated by mixed magnesium oxide and sodium hydroxide. Cement Concr. Compos. **127**, 104397 (2022)
47. A.M. Rashad, O.M. Haraz, A. Elboushi, W.M. Morsi, Electrical properties of alkali-activated materials against Portland cement. Proc. Inst. Civ. Eng. Constr. Mater **176**(1), 33–44 (2023)
48. Z. Shi, C. Shi, J. Zhang, S. Wan, Z. Zhang, Z. Ou, Alkali-silica reaction in waterglass-activated slag mortars incorporating fly ash and metakaolin. Cem. Concr. Res. **108**, 10–19 (2018)
49. H. Alanazi, J. Hu, Y.-R. Kim, Effect of slag, silica fume, and metakaolin on properties and performance of alkali-activated fly ash cured at ambient temperature. Constr. Build. Mater. **197**, 747–756 (2019)
50. M. Gómez-Casero, C. De Dios-Arana, J. Bueno-Rodríguez, L. Pérez-Villarejo, D. Eliche-Quesada, Physical, mechanical and thermal properties of metakaolin-fly ash geopolymers. Sustain. Chem. Pharm. **26**, 100620 (2022)
51. P. Duan, C. Yan, W. Zhou, Influence of partial replacement of fly ash by metakaolin on mechanical properties and microstructure of fly ash geopolymer paste exposed to sulfate attack. Ceram. Int. **42**(2), 3504–3517 (2016)
52. C. Fu, H. Ye, K. Zhu, D. Fang, J. Zhou, Alkali cation effects on chloride binding of alkali-activated fly ash and metakaolin geopolymers. Cement Concr. Compos. **114**, 103721 (2020)
53. O. Burciaga-Díaz, J.I. Escalante-García, R. Arellano-Aguilar, A. Gorokhovsky, Statistical analysis of strength development as a function of various parameters on activated metakaolin/slag cements. J. Am. Ceram. Soc. **93**(2), 541–547 (2010)
54. A.M. Rashad, A.S. Ouda, Effect of tidal zone and seawater attack on high-volume fly ash pastes enhanced with metakaolin and quartz powder in the marine environment. Microporous Mesoporous Mater. **324**, 111261 (2021)
55. A. Buchwald, H. Hilbig, C. Kaps, Alkali-activated metakaolin-slag blends—performance and structure in dependence of their composition. J. Mater. Sci. **42**(9), 3024–3032 (2007)
56. A.M. Rashad, Investigation on high-volume fly ash pastes modified with micro-size metakaolin subjected to high temperatures. J. Central South Univ. **27**(1), 231–241 (2020)
57. H. Peng, C. Cui, Z. Liu, C. Cai, Y. Liu, Synthesis and reaction mechanism of an alkali-activated metakaolin-slag composite system at room temperature. J. Mater. Civ. Eng. **31**(1), 04018345 (2019)
58. A.B. Pascual, T.M. Tognonvi, A. Tagnit-Hamou, Optimization study of waste glass powder-based alkali activated materials incorporating metakaolin: activation and curing conditions. J. Clean. Prod. **308**, 127435 (2021)
59. H.Y. Zhang, V. Kodur, B. Wu, L. Cao, S.L. Qi, Comparative thermal and mechanical performance of geopolymers derived from metakaolin and fly ash. J. Mater. Civ. Eng. **28**(2), 04015092 (2016)
60. P. Rovnaník, P. Rovnanikova, M. Vyšvařil, S. Grzeszczyk, E. Janowska-Renkas, Rheological properties and microstructure of binary waste red brick powder/metakaolin geopolymer. Constr. Build. Mater. **188**, 924–933 (2018)
61. Z. Chen, H. Ye, Improving sulphuric acid resistance of slag-based binders by magnesium-modified activator and metakaolin substitution. Cement Concr. Compos. **131**, 104605 (2022)
62. O. Burciaga-Díaz, J.I. Escalante-García, Comparative performance of alkali activated slag/metakaolin cement pastes exposed to high temperatures. Cement Concr. Compos. **84**, 157–166 (2017)
63. M. Morsy, S. Shebl, A. Rashad, Effect of fire on microstructure and mechanical properties of blended cement pastes containing metakaolin and silica fume (2008)
64. M. Morsy, A. Rashad, S. Shebl, Effect of elevated temperature on compressive strength of blended cement mortar. Build. Res. J. **56**(2–3), 173–185 (2008)

65. A.M. Rashad, D.M. Sadek, An exploratory study on alkali-activated slag blended with micro-size metakaolin particles under the effect of seawater attack and tidal zone. Arab. J. Sci. Eng. **47**(4), 4499–4510 (2022)

Chapter 2
Effect of Metakaolin on the Fresh Properties of Geopolymers

2.1 Workability

El Moustapha et al. [1] found 16.67% and 30.83% higher flowability of alkali-activated slag (AAS) mortar mixtures activated with NaOH and sodium silicate solution by partially replacing slag (fineness 445 m^2/kg) with 5% and 10% MK (fineness 17,000 m^2/kg), respectively. Huseien et al. [2] partially replaced slag (~60% of the particles were less than 10 μm) with 5–15% MK (75% of the particles were less than 10 μm) in mortar mixtures activated with NaOH and sodium silicate solution. The results indicated ~29.6%, ~37% and ~40.74% higher flowability of the mixtures with the inclusion of 5%, 10% and 15% MK, respectively, even though the fineness of MK was higher than that of slag. Khalil et al. [3] partially replaced slag (fineness 408.8 m^2/kg) with 50% MK (fineness 18,000 m^2/kg) in mortar mixtures activated with NaOH and sodium silicate solution. They found an increase in the workability with increasing SiO_2/Na_2O ratio. When the SiO_2/Na_2O ratio was 1.1, 1.3 and 1.7, the flowability of 50% slag/50% MK was 30%, 50% and 150%, respectively. Compared to AAS mixtures, the incorporation of 50% MK increased the flowability by 50% and 25% when the SiO_2/Na_2O ratio was 1.1 and 1.3, respectively. Contradictory, when the SiO_2/Na_2O ratio was 1.7, the incorporation of MK decreased the flowability by 13.33%. They increased the SiO_2/Na_2O ratio by increasing the sodium silicate ratio. Increasing the sodium silicate ratio resulted in improved workability due to more dissolving of raw materials and increasing water content. Chen and Ye [4] found 16.67% reduction in the flowability of the AAS paste mixture activated with NaOH and MgO by partially replacing slag (fineness 501 m^2/kg) with 20% MK.

Alanazi et al. [5] found ~12.88% and ~25% reduction in the flowability of fly ash (FA) mortar mixtures activated with NaOH and sodium silicate solution by partially replacing FA with 5% and 10% MK, when sodium silicate/NaOH was 1, whilst the reduction was ~19% and 27.3%, respectively, when sodium silicate/NaOH was 2.5. They related this reduction to the high surface area of MK. Abbass et al. [6] found ~1.12% and ~14.6% reduction in the flowability of FA geopolymer paste mixtures

A. M. Rashad, *Metakaolin Effect on Geopolymers' Properties*,
SpringerBriefs in Applied Sciences and Technology,
https://doi.org/10.1007/978-3-031-45151-5_2

activated with NaOH and sodium silicate solution by partially replacing FA (size 37.92 μm) by 5% and 20% MK (size 16.05 μm), respectively. When the size of MK was 2.92 μm, the reduction in the flowability was ~11.23% and ~21.3%, respectively. Bheel et al. [7] found ~9%, ~16.7%, ~20% and ~26.7% reduction in the flowability of FA self-compacting concrete mixtures activated with NaOH and sodium hydroxide solution by partially replacing FA (fineness 379 m^2/kg) with 5%, 10%, 15% and 20% MK (fineness 18,000 m^2/kg), respectively. They related this reduction to the high surface area of MK compared to that of FA. Duan et al. [8] partially replaced FA (d50 51.09 μm) with 5–30 MK (d50 12.66 μm) in paste mixtures activated with NaOH and sodium silicate solution. The results showed ~0.1%, ~8.4%, ~16.84%, 30%, 52.1% and 62.63% reduction in the flowability with the inclusion of 5%, 10%, 15%, 20%, 25% and 30% MK, respectively. They related this reduction to the higher water demand of MK due to its high surface area. Fu et al. [9] activated FA paste mixtures with NaOH or KOH. The FA (size 19.3 μm) was replaced with 20–100% MK (size 1.8 μm). When NaOH was used, the incorporation of 20%, 50%, 80% and 100% MK increased the flow by 8.33%, 6.67%, 7.78% and 16.67%, respectively, when KOH was used, the flow was decreased by 6.82%, 7.73%, 2.27% and 0%, respectively.

Sharmin et al. [10] found 14.28%, 57.14%, 100% and 114.28% increase in the flow of 40% rice husk ash (RHA)/60% slag mortar mixtures activated with NaOH and sodium silicate solution by partially replacing slag (fineness 405 m^2/kg) with 5%, 10%, 15% and 20% MK (fineness 4315.8 m^2/kg), respectively. Furlani et al. [11] found lower workability of steel slag paste mixtures activated with NaOH and sodium silicate solution by partially replacing steel slag (d50 36 or 70 μm) with 20–100% MK (size < 40 μm). As the amount of MK increased, the workability decreased. Rovnaník et al. [12] found higher workability of waste brick powder (WBP) geopolymer paste mixtures activated with NaOH and sodium silicate solution by replacing WBP (d50 = 8.5 μm) with 25–100% MK (d50 = 6.3 μm). The workability results are summarized in Table 2.1.

Considering what has been reported thus far, it can be confirmed that the incorporation of MK into geopolymer mixtures may have a negative or a positive effect on workability. This mainly depends on the type of the main precursor. Other factors affecting workability are SiO_2/Na_2O ratio and activator type. However, when slag was used as the main precursor, the incorporation of MK increased the workability (Fig. 2.1) even though its particle size was smaller than slag. Huseien et al. [2] related this improvement in the workability to the reduction in the number of angular slag particles and calcium content by including MK. The incorporation of MK led to an increase in aluminium and silicate. Increasing silicate content led to improving the flowability [13]. Similarly, when WBP was used as the main precursor, the incorporation of MK increased the workability. Rovnaník et al. [12] related this improvement to the rapid MK dissolution in the activator solution and the consequent increase in the amount of free water due to SiO_4 and AlO_4 condensation. Contrarily, when FA was used as the main precursor, the incorporation of MK decreased the workability (Fig. 2.1). This negative effect of MK on the workability is attributed to its irregular spiny plate-like particle compared to the spherical particle of FA. In

Table 2.1 Effect of MK on the mixture workability

References	Precursor	Precursor size/ surface area	Activator	MK (%)	MK size/ surface area	Type	Effect	Ratio
El Moustapha et al. [1]	Slag	445 m²/kg	NaOH and sodium silicate	5 and 10	17,000 m²/kg	Mortar	Increased	16.67 and 30.83
Huseien et al. [2]	Slag	60% < 10 μm	NaOH and sodium silicate solution	5, 10 and 15	75% < 10 μm	Mortar	Increased	~29.6, ~37 and ~40.74
Khalil et al. [3]	Slag	408.8 m²/kg	NaOH and sodium silicate solution	50	18,000 m²/kg	Mortar	Increased	30, 50 and 150 (according to SiO_2/Na_2O ratio)
Chen and Ye [4]	Slag	501 m²/kg	NaOH and MgO	20	> 501 m²/kg	Paste	Decreased	16.67
Alanazi et al. [5]	FA	–	NaOH and sodium silicate	5 and 10	–	Mortar	Decreased	~12.88 and ~25
Abbass et al. [6]	FA	37.92	NaOH and sodium silicate	5 and 20	16.05	Paste	Decreased	~1.12 and ~14.6
	FA	37.92	NaOH and sodium silicate	5 and 20	2.92	Paste	Decreased	~11.23 and ~21.3
Bheel et al. [7]	FA	379 m²/kg	NaOH and sodium silicate solution	5, 10, 15 and 20	18,000 m²/kg	Concrete	Decreased	~9, ~16.7, ~20 and ~26.7
Duan et al. [8]	FA	51.09 μm	NaOH and sodium silicate	5, 10, 15, 20, 25 and 30	12.66 μm	Paste	Decreased	~0.1, ~8.4, ~16.84, 30, 52.1 and 62.63
Fu et al. [9]	FA	19.3 μm	KOH	20, 50, 80 and 100	1.8 μm	Paste	Decreased	6.82, 7.73, 2.27 and 0
	FA	19.3 μm	NaOH	20, 50, 80 and 100	1.8 μm	Paste	Increased	8.33, 6.67, 7.78 and 16.67
Sharmin et al. [10]	RHA/slag	2981.2/405 m²/kg	NaOH and sodium silicate	5, 10, 15 and 20	4315.8 m²/kg	Mortar	Increased	14.28, 57.14, 100 and 114.28

(continued)

Table 2.1 (continued)

References	Precursor	Precursor size/ surface area	Activator	MK (%)	MK size/ surface area	Type	Effect	Ratio
Furlani et al. [11]	Steel slag	d50 36 or 70 μm	NaOH and sodium silicate	20–100	< 40 μm	Paste	Decreased	–
Rovnaník et al. [12]	WBP	d50 8.5 μm	NaOH and sodium silicate	25–100	d50 6.3 μm	Paste	Increased	–

addition to the higher water demand of MK due to its high fineness [5, 7, 8]. In actuality, around 8.33%, 41.67% and 50% of the quoted references focused on concrete, mortar and paste mixtures, respectively (Fig. 2.2), whilst 33.33%, 41.67% and 25% of these references used slag precursor, FA precursor and other precursors, respectively (Fig. 2.3).

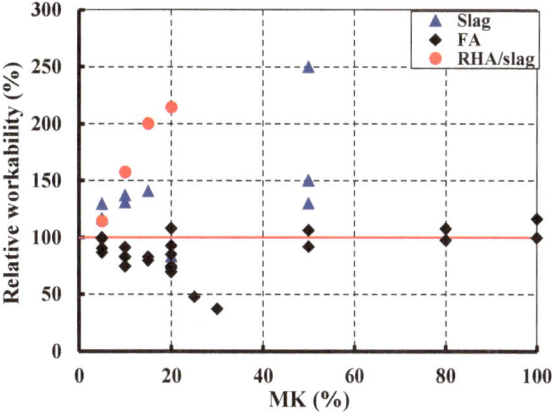

Fig. 2.1 Effect of MK on the workability of different types of geopolymer [1–12]

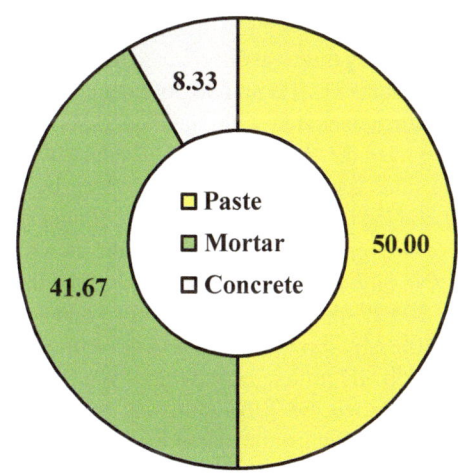

Fig. 2.2 Percentage of workability studies of various media containing MK

Fig. 2.3 Percentage of research of various precursors containing MK for workability

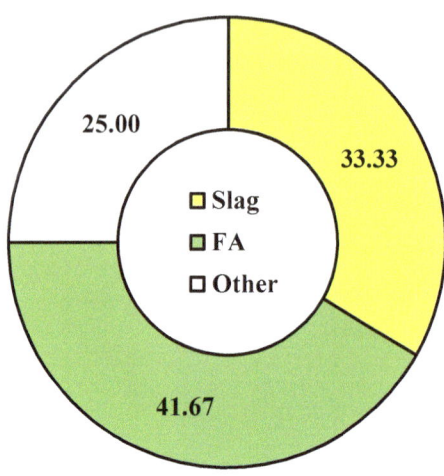

2.2   Setting Time

Huseien et al. [2] found a deceleration of the initial and final setting time of AAS mortar mixtures activated with NaOH and sodium silicate solution by partially replacing slag with 5–15% MK. The incorporation of 5%, 10% and 15% MK decelerated the initial setting time by ~169%, ~236% and ~284%, respectively, whilst the final setting time was decelerated by ~133%, ~166% and ~216%, respectively. Li et al. [14] found 21.21% and 36.36% deceleration of the initial setting time of AAS paste mixtures activated with NaOH and sodium silicate solution by partially replacing slag (size 0.1–50 μm) with 10% and 20% MK (0.15–142 μm), respectively, whilst the final setting time was decelerated by 38.98% and 44.1%, respectively. Bernal et al. [15] found 59.14% and 50% longer initial and final setting time of AAS paste mixtures activated with sodium silicate with the incorporation of 10% MK when the Si/Na ratio was 2.4, respectively, whilst the incorporation of 20% MK prolonged it by 203.2% and 196.7%, respectively. When the Si/Na ratio was 2, the incorporation of 10% MK showed a comparable setting time to the control, whilst the incorporation of 20% MK prolonged the initial and final setting time by 1.39% and 105%, respectively. When the Si/Na ratio was 1.6, the incorporation of 10% MK prolonged the initial and final setting time by 130.7% and 104.5%, respectively, whilst the incorporation of 20% MK prolonged it by 192.3% and 163.6%, respectively. Bernal [16] found a longer setting time of AAS paste mixtures activated with NaOH and sodium silicate solution when slag (fineness 399 m^2/kg) was partially replaced with 10% and 20% MK (fineness 391 m^2/kg). When the SiO$_2$/Al$_2$O$_3$ was 4.4, the incorporation of 10% MK decelerated the initial and final setting time by 17.39% and 54.28%, respectively, whilst the incorporation of 20% MK decelerated it by 65.22% and 80%, respectively. When the SiO$_2$/Al$_2$O$_3$ was 4, the incorporation of 10% MK decelerated the initial and final setting time by 318.18% and 133.33%, respectively, whilst the incorporation of 20% MK decelerated it by 209.1% and 120%, respectively. When the SiO_2/Al_2O_3 was 3.6, the incorporation of 10% MK decelerated the initial and final setting time by 240% and 42.85%, respectively, whilst the incorporation of 20% MK decelerated it by 180% and 28.57%, respectively. Peng et al. [17] prepared AAS pastes activated with NaOH and sodium silicate solution. The slag (BET fineness 15 m^2/g) was replaced with 20–100% MK (BET fineness 20 m^2/g). Different Si/Na ratios of 1.4–2 were used. The results showed longer setting time with the inclusion of MK. Li et al. [18] found 12.82% and 13.67% longer initial setting time of 50% slag/50% FA pastes activated with NaOH and sodium silicate solution by partially replacing slag with 5% and 10% MK, respectively, whilst the final setting time was prolonged by 21.71% and 27.43%, respectively. They related this deceleration to the decreased reaction rate, which decreased the reaction products, with the inclusion of MK. Guo et al. [19] found a longer setting time of FA paste mixtures activated with phosphoric acid by partially replacing FA (fineness 8.16 m^2/cm^3) with 20%, 30% and 50% MK (fineness 0.77 m^2/cm^3). The incorporation of 20%, 30% and 50% MK prolonged the initial setting time by ~1.6, ~8.3 and ~75 folds, respectively, whilst the final setting time was prolonged by ~8, ~24 and ~56 folds, respectively. They related this deceleration to the slow process of polycondensation of matrix with including MK. On the other hand, Zhang et al. [20] found shorter initial and final setting time of the composites of FA/MK with increasing MK ratios from 10% to 90%. The results included in this section are summarized in Table 2.2.

Considering what has been reported thus far, it can be noticed that all studies agreed that the incorporation of MK in the mixture decelerated the setting time regardless the precursor type. The rate of deceleration depends on precursor type, Si/Na, Si/Al and MK amount. When FA was used as the main precursor, the rate of deceleration of setting time with including MK was higher than that of the slag precursor. When slag was used as the main precursor, the incorporation of MK decelerated setting time due to the reduction in the calcium content and increment in Al_2O_3 and SiO_2 with including MK [2]. When other precursors were used, the incorporation of MK decelerated setting time due to the slow process of polycondensation of the matrix [19] and the decreased reaction rate [18] with including MK. In actuality, around 50% and 50% of the quoted references focused on paste and mortar mixtures, respectively (Fig. 2.4), whilst 50%, 33.33% and 16.67% of these references used slag precursor, FA precursor and slag/FA precursor, respectively (Fig. 2.5).

Table 2.2 Effect of MK on the setting time of geopolymer mixtures

References	Precursor	Precursor size/surface area	Activator	MK (%)	MK size/surface area	Type	Effect	Ratio
Huseien et al. [2]	Slag	60% < 10 μm	NaOH and sodium silicate solution	5, 10 and 15	75% < 10 μm	Mortar	Decelerated initial setting time	~169, ~236 and ~284
	Slag	60% < 10 μm	NaOH and sodium silicate solution	5, 10 and 15	75% < 10 μm	Mortar	Decelerated final setting time	~133, ~166 and ~216
Li et al. [14]	Slag	0.1–50 μm	NaOH and sodium silicate	10	0.15–142 μm	Paste	Decelerated initial setting time	21.21 and 36.36
	Slag	0.1–50 μm	NaOH and sodium silicate	10	0.15–142 μm	Paste	Decelerated final setting time	38.98 and 44.1
Bernal et al. [15]	Slag	0.1–74 μm	Sodium silicate	10	1.8–100 μm	Mortar	Decelerated initial and final setting time	59.14 and 50 Si/Na = 2.4)
	Slag	0.1–74 μm	Sodium silicate	20	1.8–100 μm	Mortar	Decelerated initial and final setting time	203.2 and 196.7 Si/Na = 2.4)
	Slag	0.1–74 μm	Sodium silicate	10	1.8–100 μm	Mortar	Comparable	~ (Si/Na = 2)
	Slag	0.1–74 μm	Sodium silicate	20	1.8–100 μm	Mortar	Decelerated initial and final setting time	1.39 and 105 (Si/Na = 2)
	Slag	0.1–74 μm	Sodium silicate	10	1.8–100 μm	Mortar	Decelerated initial and final setting time	130.7 and 104.5 (Si/Na = 1.6)
	Slag	0.1–74 μm	Sodium silicate	20	1.8–100 μm	Mortar	Decelerated initial and final setting time	192.3 and 163.6 Si/Na = 1.6)

(continued)

Table 2.2 (continued)

References	Precursor	Precursor size/surface area	Activator	MK (%)	MK size/surface area	Type	Effect	Ratio
Bernal [16]	Slag	399 m^2/kg	NaOH and sodium silicate	10	391 m^2/kg	Mortar	Decelerated initial and final setting time	17.39 and 54.28 (SiO$_2$/Al$_2$O$_3$ = 4.4)
	Slag	399 m^2/kg	NaOH and sodium silicate	20	391 m^2/kg	Mortar	Decelerated initial and final setting time	65.22 and 80 (SiO$_2$/Al$_2$O$_3$ = 4.4)
	Slag	399 m^2/kg	NaOH and sodium silicate	10	391 m^2/kg	Mortar	Decelerated initial and final setting time	318.18 and 133.33 (SiO$_2$/Al$_2$O$_3$ = 4)
	Slag	399 m^2/kg	NaOH and sodium silicate	20	391 m^2/kg	Mortar	Decelerated initial and final setting time	209.1 and 120 (SiO$_2$/Al$_2$O$_3$ = 4)
	Slag	399 m^2/kg	NaOH and sodium silicate	10	391 m^2/kg	Mortar	Decelerated initial and final setting time	240% and 42.85 (SiO$_2$/Al$_2$O$_3$ = 3.6)
	Slag	399 m^2/kg	NaOH and sodium silicate	20	391 m^2/kg	Mortar	Decelerated initial and final setting time	180% and 28.57 (SiO$_2$/Al$_2$O$_3$ = 3.6)
Li et al. [18]	Slag/FA	18.3 µm/48.1 µm	NaOH and sodium silicate	5 and 10	69.4 µm	Paste	Decelerated initial setting time	12.82 and 13.67
	Slag/FA	18.3 µm/48.1 µm	NaOH and sodium silicate	5 and 10	69.4 µm	Paste	Decelerated final setting time	21.71 and 27.43
Guo et al. [19]	FA	D90: 3.34 µm	Phosphoric acid	20, 30 and 50	D90: 58.35 µm	Paste	Decelerated initial setting time	~60, ~730 and ~7400

(continued)

Table 2.2 (continued)

References	Precursor	Precursor size/ surface area	Activator	MK (%)	MK size/surface area	Type	Effect	Ratio
	FA	D90: 3.34 μm	Phosphoric acid	20, 30 and 50	D90: 58.35 μm	Paste	Decelerated final setting time	~700, ~2300 and ~5500
Zhang et al. [20]	FA	37 μm	NaOH and waterglass	10–90	8 μm	Paste	Shortened setting time	–

Fig. 2.4 Percentage of
setting time studies of
various media containing
MK

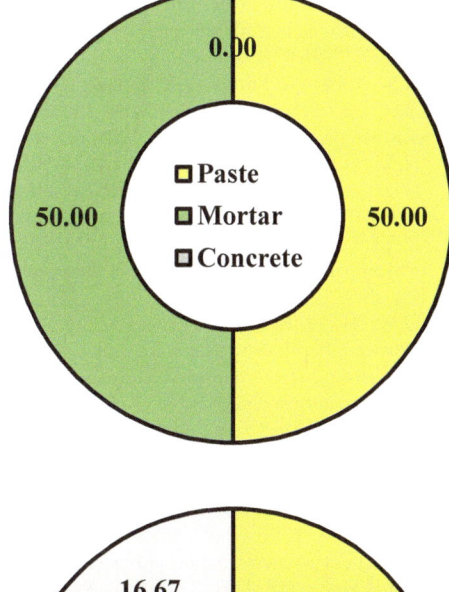

Fig. 2.5 Percentage of
research of various
precursors containing MK
for setting time

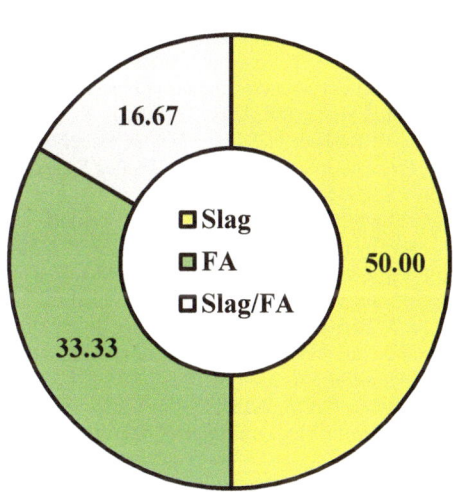

References

1. B. El Moustapha, S. Bonnet, A. Khelidj, N. Leklou, D. Froelich, I.A. Babah, C. Charbuillet, A. Khalifa, Compensation of the negative effects of micro-encapsulated phase change materials by incorporating metakaolin in geopolymers based on blast furnace slag. Constr. Build. Mater. **314**, 125556 (2022)
2. G.F. Huseien, J. Mirza, M. Ismail, S. Ghoshal, M.A.M. Ariffin, Effect of metakaolin replaced granulated blast furnace slag on fresh and early strength properties of geopolymer mortar. Ain Shams Eng. J. **9**(4), 1557–1566 (2018)
3. M.G. Khalil, F. Elgabbas, M.S. El-Feky, H. El-Shafie, Performance of geopolymer mortar cured under ambient temperature. Constr. Build. Mater. **242**, 118090 (2020)
4. Z. Chen, H. Ye, Improving sulphuric acid resistance of slag-based binders by magnesium-modified activator and metakaolin substitution. Cement Concr. Compos. **131**, 104605 (2022)

5. H. Alanazi, J. Hu, Y.-R. Kim, Effect of slag, silica fume, and metakaolin on properties and performance of alkali-activated fly ash cured at ambient temperature. Constr. Build. Mater. **197**, 747–756 (2019)
6. A.M. Abbass, R. Firdous, J.N. Yankwa Djobo, D. Stephan, M.A. Elrahman, The role of chemistry and fineness of metakaolin on the fresh properties and heat resistance of blended fly ash-based geopolymer. SN Appl. Sci. **5**(5) 1–17 (2023)
7. N. Bheel, P. Awoyera, T. Tafsirojjaman, N.H. Sor, Synergic effect of metakaolin and groundnut shell ash on the behavior of fly ash-based self-compacting geopolymer concrete. Constr. Build. Mater. **311**, 125327 (2021)
8. P. Duan, C. Yan, W. Zhou, Influence of partial replacement of fly ash by metakaolin on mechanical properties and microstructure of fly ash geopolymer paste exposed to sulfate attack. Ceram. Int. **42**(2), 3504–3517 (2016)
9. C. Fu, H. Ye, K. Zhu, D. Fang, J. Zhou, Alkali cation effects on chloride binding of alkali-activated fly ash and metakaolin geopolymers. Cement Concr. Compos. **114**, 103721 (2020)
10. A. Sharmin, U.J. Alengaram, M.Z. Jumaat, M.O. Yusuf, S.A. Kabir, I.I. Bashar, Influence of source materials and the role of oxide composition on the performance of ternary blended sustainable geopolymer mortar. Constr. Build. Mater. **144**, 608–623 (2017)
11. E. Furlani, S. Maschio, M. Magnan, E. Aneggi, F. Andreatta, M. Lekka, A. Lanzutti, L. Fedrizzi, Synthesis and characterization of geopolymers containing blends of unprocessed steel slag and metakaolin: the role of slag particle size. Ceram. Int. **44**(5), 5226–5232 (2018)
12. P. Rovnaník, P. Rovnanikova, M. Vyšvařil, S. Grzeszczyk, E. Janowska-Renkas, Rheological properties and microstructure of binary waste red brick powder/metakaolin geopolymer. Constr. Build. Mater. **188**, 924–933 (2018)
13. A.M. Rashad, in *Silica Fume in Geopolymers: A Comprehensive Review of Its Effects on Properties. SpringerBriefs in Applied Science and Technology* (2023), https://doi.org/10.1007/978-3-031-33219-7.
14. Z. Li, M. Nedeljković, B. Chen, G. Ye, Mitigating the autogenous shrinkage of alkali-activated slag by metakaolin. Cem. Concr. Res. **122**, 30–41 (2019)
15. S.A. Bernal, J.L. Provis, V. Rose, R.M. De Gutierrez, Evolution of binder structure in sodium silicate-activated slag-metakaolin blends. Cement Concr. Compos. **33**(1), 46–54 (2011)
16. S.A. Bernal, Effect of the activator dose on the compressive strength and accelerated carbonation resistance of alkali silicate-activated slag/metakaolin blended materials. Constr. Build. Mater. **98**, 217–226 (2015)
17. H. Peng, C. Cui, Z. Liu, C. Cai, Y. Liu, Synthesis and reaction mechanism of an alkali-activated metakaolin-slag composite system at room temperature. J. Mater. Civ. Eng. **31**(1), 04018345 (2019)
18. Z. Li, X. Liang, Y. Chen, G. Ye, Effect of metakaolin on the autogenous shrinkage of alkali-activated slag-fly ash paste. Constr. Build. Mater. **278**, 122397 (2021)
19. H. Guo, B. Zhang, L. Deng, P. Yuan, M. Li, Q. Wang, Preparation of high-performance silico-aluminophosphate geopolymers using fly ash and metakaolin as raw materials. Appl. Clay Sci. **204**, 106019 (2021)
20. X. Zhang, X. Zhang, X. Li, M. Ma, Z. Zhang, X. Ji, Slurry rheological behaviors and effects on the pore evolution of fly ash/metakaolin-based geopolymer foams in chemical foaming system with high foam content. Constr. Build. Mater. **379**, 131259 (2023)

Chapter 3
Effect of Metakaolin on the Physical Properties of Geopolymers

3.1 Density

El Moustapha et al. [1] found 0.95% and 1.85% higher 28 days apparent density of alkali-activated slag (AAS) mortars activated with NaOH and sodium silicate solution by partially replacing slag with 5% and 10% MK, respectively. Huseien et al. [2] found ~1.7%, ~4.8% and 5.5% higher density of AAS mortars activated with NaOH and sodium silicate solution by partially replacing slag with 5%, 10% and 15% MK, respectively. Gharieb et al. [3] found 13.4% and 5.3% increase in the 28 days bulk density of AAS pastes activated with NaOH and sodium silicate solution cured in air and water, respectively, by partially replacing slag (average diameter = 37.183 μm) with 30% MK (average diameter = 4.507 μm). Gómez-Casero et al. [4] prepared fly ash (FA) pastes activated with NaOH and sodium silicate solution. The FA (d50 19.87 μm) was replaced with 25–100% MK (d50 9.579 μm). The specimens were cured at 60 °C with standard humidity for one day. The results showed that the 28 days bulk density slightly increased with increasing MK content. Zhang et al. [5] found 78.5% and 9.7% increment in the apparent density of FA geopolymer pastes activated with KOH and potassium silicate solution by replacing FA (size 32 μm) with 50% and 100% MK (size 17 μm), respectively. Sharmin et al. [6] found 12.59%, 13.3%, 13.7% and 14.31% higher density of 40% rice husk ash (RHA)/60% slag geopolymer mortars activated with NaOH and sodium silicate solution by partially replacing slag with 5%, 10%, 15% and 20% MK, respectively. Contrarily, Rajamma et al. [7] found 10.36% and 11.38% reduction in the 10 days density of biomass FA geopolymer pastes activated with NaOH and sodium silicate solution by partially replacing FA with 20% and 40% MK, respectively, when 10 M of NaOH was used, whilst the reduction reached 16.18% and 14.45%, respectively, when 18 M of NaOH was used. Bignozzi et al. [8] found 10.53%, 17.37%, 17.9%, 18.4%, 25.26% and 32.63% reduction in the 7 days bulk density of ladle slag geopolymer pastes with the incorporation of 20%, 30%, 40%, 50%, 75% and 100% MK, respectively.

© The Author(s), under exclusive license to Springer Nature Switzerland AG 2024   25
A. M. Rashad, *Metakaolin Effect on Geopolymers' Properties*,
SpringerBriefs in Applied Sciences and Technology,
https://doi.org/10.1007/978-3-031-45151-5_3

Considering what has been reported thus far, it can be noticed that the variation in the density with the incorporation of MK mainly depends on precursor type and activator concentration. When slag or RHA/slag was used as the main precursor, the incorporation of MK increased the density due to the formation of N–A–S–H gel with including SiO_2 and Al_2O_3 from MK beside C–S–H gel, which can enhance the microstructure [2]. El Moustapha et al. [1] related this higher density to the high reactivity of MK compared to slag. This high reactivity caused the good dissolution of alumina and silica to form N–A–S–H gel and C–(A)–S–H gel. When FA was used as the main precursor, the increased density with including MK could be related to the higher density of MK (2569 kg/m^3) compared to that of FA (2480 kg/m^3) [4]. As opposed to that, the incorporation of MK into ladle slag geopolymers decreased the density due to the reduction in the specific gravity.

3.2 Porosity and Water Absorption

El Moustapha et al. [1] found 6.2% and 11.63% higher 28 days porosity of AAS mortars activated with NaOH and sodium silicate solution by partially replacing slag with 5% and 10% MK, respectively. Bernal et al. [9] prepared AAS concrete specimens activated with NaOH and sodium silicate solution. The slag (size 0.1–74 μm d50 15 μm) was partially replaced with 10% and 20% MK (size 1.8–100 μm d50 12.2 μm). Different ratios of Si/Al were used. The specimens were cured at 25 ± 5 °C with 90% RH. When the Si/Al ratio of 3.6 was used, the 28 days water absorption was decreased by ~20% with the inclusion of 10% MK, whilst the inclusion of 20% MK increased it by ~10%. The 90 days water absorption was increased by ~42.85% and ~71.4% with the inclusion of 10% and 20% MK, respectively. When the Si/Al ratio of 4 was used, the 28 days water absorption was decreased by ~50% with the inclusion of 10% MK, whilst the inclusion of 20% MK increased it by ~16.7%. The 90 days water absorption was decreased by ~60% and ~50% with the inclusion of 10% and 20% MK, respectively. When the Si/Al ratio of 4.4 was used, the 28 days water absorption was increased by 11.5% with the inclusion of 10% MK, whilst the inclusion of 20% MK decreased it by ~44.2%. The 90 days water absorption was decreased by ~13.3% and ~33.3% with the inclusion of 10% and 20% MK, respectively. When the Si/Al ratio of 3.6 was used, the 28 days total porosity was decreased by ~23.8% and ~38.5% with the inclusion of 10% and 20% MK, respectively, whilst the 90 days porosity was reduced by ~17.9% and 37.9%, respectively. When the Si/Al ratio of 4 was used, the 28 days porosity was decreased by ~65.25% and ~48.3% with the incorporation of 10% and 20% MK, respectively, whilst the 90 days porosity was decreased by ~40.3% and ~41.9%, respectively. When a Si/Al ratio of 4.4 was used, the 28 days porosity was decreased by ~42.1% and ~5.3% with the inclusion of 10% and 20% MK, respectively, whilst the 90 days porosity was decreased by ~20% and ~37.1%, respectively. El Moustapha et al. [10] found 5.38% and 10.77% higher 28 days water absorption of AAS mortars activated with NaOH and sodium silicate with the incorporation of 10% and 20%

MK, respectively, whilst the capillary water absorption was increased by 2.13% and 6.38%, respectively. The MIP porosity was decreased by 66.67% and 16.67% with the incorporation of 10% and 20% MK, respectively. Asaad et al. [11] found ~1.6% and ~11.29% reduction in the 28 days porosity of AAS mortars activated with NaOH and sodium silicate solution with the incorporation of 5% and 10% MK, respectively, whilst the incorporation of 15% and 25% MK increased it by ~3.2% and ~11.29%, respectively. At the age of 180 days, the incorporation of 10% MK decreased the porosity by 12.96%. At the age of 360 days, the incorporation of 25% MK increased the porosity by 1.88%. Gharieb et al. [3] found 15.4% and 18.1% reduction in the 28 days total porosity of AAS pastes activated with NaOH and sodium silicate solution by partially replacing slag (average diameter = 37.183 μm) with 30% MK (average diameter = 4.507 μm) when air curing and water curing were used, respectively. Bernal et al. [12] found 4.8%, 44%, 54.53% and 69.63% higher 28 days permeable volumes of pores of AAS mortars activated with NaOH and sodium silicate solution by partially replacing slag (fineness 498 m^2/kg) with 20%, 40%, 60% and 80% MK (fineness 391 m^2/kg), respectively.

Alanazi et al. [13] found 11.69% reduction in the 28 days porosity of 90% FA/ 10% MK mortars activated with NaOH and sodium silicate solution cured at room temperature in comparison with those of 100% FA cured at 60 °C for 24 h. Bheel et al. [14] found 60.95%, 51.2%, 33.33% and 19.04% reduction in the water penetration depth of FA self-compacting concrete specimens activated with NaOH and sodium silicate solution by partially replacing FA with 20%, 15%, 10% and 5% MK, respectively. Abbass et al. [15] found a reduction in the 56 days apparent porosity and water absorption of FA geopolymer pastes activated with NaOH and sodium silicate solution by partially replacing FA with 5% and 20% MK. Zhang et al. [16] prepared foamed geopolymer pastes from 0.9/0.1, 0.7/0.3, 0.5/0.5, 0.3/0.7 and 0.1/ 0.9 FA/MK activated with NaOH and waterglass solution. The results of porosity showed an increasing trend with increasing MK ratio. The specimens prepared from 0.9/0.1, 0.7/0.3, 0.5/0.5, 0.3/0.7 and 0.1/0.9 showed 28 days porosity of ~83.25%, 84.75%, ~85%, ~85.6% and 86%, respectively. Compared to 0.9/0.1 specimens, the porosity of the 0.7/0.3, 0.5/0.5, 0.3/0.7 and 0.1/0.9 increased by ~1.8%, ~2.1%, ~2.8 and ~3.3%, respectively. Zhang et al. [5] found 8.9% and 14.7% reduction in the MIP porosity of FA geopolymer pastes activated with KOH and potassium silicate solution by replacing FA (size 32 μm) with 50% and 100% MK (size 17 μm), respectively. Gómez-Casero et al. [4] found ~16%, ~20.5%, ~42.8% and 59.82% higher apparent porosity of FA pastes activated with NaOH and sodium silicate solution by replacing FA with 25%, 50%, 75% and 100% MK, respectively, whilst the water absorption was increased by ~25%, ~42.58, ~57.14% and 67.14%, respectively. The incorporation of 25%, 50%, 75% MK decreased the 28 days apparent porosity by ~0.8%, 11.4%, ~13.5% and 19.8%, respectively, whilst the water absorption was decreased by ~0.7%, ~2.7%, ~7% and 7.4%, respectively. Bignozzi et al. [8] found 21.6% lower 7 days total open porosity of ladle slag geopolymer pastes activated with NaOH and sodium silicate solution by partially replacing ladle slag (95% passing at 90 μm) with 20% MK (95% passing at 30 μm). However, the incorporation of 30%, 40%, 50%, 75% and 100% MK increased the porosity by 4.98%, 5.65%, 14.29%,

28.24% and 41.86%, respectively. The results included in this part are summarized in Table 3.1.

Considering what has been reported thus far, it is clear that the findings on the effect of MK on geopolymers porosity and water absorption are inconclusive (Figs. 3.1 and 3.2), but most of them (around 38.46%) confirmed an improvement (reduction) of porosity and water absorption with including MK, whilst around 30.77% of them confirmed adversely affect. The remaining studies (around 30.77%) reported that the incorporation of MK may decrease or increase the porosity and water absorption (Fig. 3.3). This typically depends on testing age, activator type/concentration, Si/Al ratio, MK ratio/fineness, precursor type/fineness and curing condition. When slag was used as the main precursor, around 63.6% of the obtained results reported higher porosity with including MK, whilst only around 36.4% of the obtained results reported lower porosity. To meet this reduction, it is recommended that the ratio of MK not exceed 10%, of which it was reported that the incorporation of high ratios of MK resulted in the low formation of C–(A)–S–H gel. The increased alumina and silica ratios than the optimum influenced the partially and non-reacted particles [17]. When FA was used as the main precursor, around 36.4% of the obtained results reported higher porosity with including MK, whilst around 63.6% of the obtained results reported lower porosity. This depends mainly on the testing age. However, the general trend is lower porosity with a lower ratio of MK. This improvement could be attributed to the MK fineness that can fill the pores [14]. When ladle slag was used as the main precursor, the incorporation of MK showed an adverse effect on porosity. As shown in Fig. 3.2, the incorporation of 10% and 20% MK into AAS concrete may increase or decrease water absorption. This mainly depends on the testing age and Si/Al ratio. When FA was used as the main precursor, it was better to not increase the ratio of the MK by more than 20% to obtain low water absorption. In actually, the mentioned references in this section that focused on paste, concrete and mortar are roughly 46.15%, 15.38% and 38.46%, respectively (Fig. 3.4).

3.3 Shrinkage

Li et al. [18] prepared pastes from 50% slag/50% FA activated with NaOH and sodium silicate solution. The slag was partially replaced by 5% and 10% MK. The results showed a reduction in autogenous shrinkage with the inclusion of MK. The higher the MK is, the lower the autogenous shrinkage is. The incorporation of 10% MK decreased the 1 day autogenous shrinkage by 65%, whilst the 7 days autogenous shrinkage was reduced by 24%. They suggested that the autogenous shrinkage consisted of two parts named self-desiccation and internal RH. The incorporation of MK can reduce the self-desiccation of the matrix due to increase porosity and reduce chemical shrinkage. Li et al. [19] found 44% and 38% lower 1 and 7 days autogenous shrinkage of AAS pastes activated with NaOH and sodium silicate solution by partially replacing slag with 10% MK, respectively, whilst partially replacing slag with 20% MK led to > 60% and about 50% reduction, respectively. Asaad et al.

Table 3.1 Effect of MK on the porosity and water absorption of geopolymers

References	Precursor	Precursor size/fineness	Activator	MK (%)	MK size/fineness	Type	Age (day)	Curing	Effect	Ratio
El Moustapha et al. [1]	Slag	445 m²/kg	NaOH and sodium silicate	5 and 10	17,000 m²/kg	Mortar	28	20 °C with 50% RH	Increased porosity	6.2 and 11.63
Bernal et al. [9]	Slag	0.1–74 μm	NaOH and sodium silicate	10	1.8–100 μm	Concrete	28	25 ± 5 °C with 90% RH	Decreased water absorption	~20 (Si/Al = 3.6)
	Slag	0.1–74 μm	NaOH and sodium silicate	20	1.8–100 μm	Concrete	28	25 ± 5 °C with 90% RH	Increased water absorption	~10 (Si/Al = 3.6)
	Slag	0.1–74 μm	NaOH and sodium silicate	10 and 20	1.8–100 μm	Concrete	90	25 ± 5 °C with 90% RH	Increased water absorption	~42.85 and ~71.4 (Si/Al = 3.6)
	Slag	0.1–74 μm	NaOH and sodium silicate	10	1.8–100 μm	Concrete	28	25 ± 5 °C with 90% RH	Decreased water absorption	~50 (Si/Na = 4)
	Slag	0.1–74 μm	NaOH and sodium silicate	20	1.8–100 μm	Concrete	28	25 ± 5 °C with 90% RH	Increased water absorption	~16.7 (Si/Al = 4)
	Slag	0.1–74 μm	NaOH and sodium silicate	10 and 20	1.8–100 μm	Concrete	90	25 ± 5 °C with 90% RH	Decreased water absorption	~60% and ~50 (Si/Al = 4)
	Slag	0.1–74 μm	NaOH and sodium silicate	10	1.8–100 μm	Concrete	28	25 ± 5 °C with 90% RH	Increased water absorption	11.5 (Si/Al = 4.4)

(continued)

Table 3.1 (continued)

References	Precursor	Precursor size/fineness	Activator	MK (%)	MK size/ fineness	Type	Age (day)	Curing	Effect	Ratio
	Slag	0.1–74 μm	NaOH and sodium silicate	20	1.8–100 μm	Concrete	28	25 ± 5 °C with 90% RH	Decreased water absorption	~44.2 (Si/Al = 4.4)
	Slag	0.1–74 μm	NaOH and sodium silicate	10 and 20	1.8–100 μm	Concrete	90	25 ± 5 °C with 90% RH	Decreased water absorption	~13.3 and ~33.3 (Si/Al = 4.4)
	Slag	0.1–74 μm	NaOH and sodium silicate	10 and 20	1.8–100 μm	Concrete	28	25 ± 5 °C with 90% RH	Decreased porosity	~23.8 and ~38.5 (Si/Al = 3.6)
	Slag	0.1–74 μm	NaOH and sodium silicate	10 and 20	1.8–100 μm	Concrete	90	25 ± 5 °C with 90% RH	Decreased porosity	~17.9 and 37.9 (Si/Al = 3.6)
	Slag	0.1–74 μm	NaOH and sodium silicate	10 and 20	1.8–100 μm	Concrete	28	25 ± 5 °C with 90% RH	Decreased porosity	~65.25% and ~48.3 (Si/Al = 4)
	Slag	0.1–74 μm	NaOH and sodium silicate	10 and 20	1.8–100 μm	Concrete	90	25 ± 5 °C with 90% RH	Decreased porosity	~40.3 and ~41.9 (Si/Al = 4)
	Slag	0.1–74 μm	NaOH and sodium silicate	10 and 20	1.8–100 μm	Concrete	28	25 ± 5 °C with 90% RH	Decreased porosity	~42.1 and ~5.3 (Si/Al = 4.4)
	Slag	0.1–74 μm	NaOH and sodium silicate	10 and 20	1.8–100 μm	Concrete	90	25 ± 5 °C with 90% RH	Decreased porosity	~20 and ~37.1 (Si/Al = 4.4)

(continued)

Table 3.1 (continued)

References	Precursor	Precursor size/fineness	Activator	MK (%)	MK size/fineness	Type	Age (day)	Curing	Effect	Ratio
El Moustapha et al. [10]	Slag	440 m²/kg	NaOH and sodium silicate	10 and 20	170 m²/kg	Mortar	28	20 °C with 50% RH	Increased water absorption	5.38 and 10.77
	Slag	440 m²/kg	NaOH and sodium silicate	10 and 20	170 m²/kg	Mortar	28	20 °C with 50% RH	Increased capillary water absorption	2.13 and 6.38
	Slag	440 m²/kg	NaOH and sodium silicate	10 and 20	170 m²/kg	Mortar	28	20 °C with 50% RH	Decreased MIP porosity	66.67 and 16.67
Asaad et al. [11]	Slag	12 μm	NaOH and sodium silicate	5 and 10	12 μm	Mortar	28	27 °C with 75% RH	Decreased porosity	~1.6 and ~11.29
	Slag	12 μm	NaOH and sodium silicate	15 and 25	12 μm	Mortar	28	27 °C with 75% RH	Increased porosity	~3.2 and ~11.29
	Slag	12 μm	NaOH and sodium silicate	10	12 μm	Mortar	180	27 °C with 75% RH	Decreased porosity	12.96
	Slag	12 μm	NaOH and sodium silicate	25	12 μm	Mortar	360	27 °C with 75% RH	Increased porosity	1.88
Gharieb et al. [3]	Slag	37.183 μm	NaOH and sodium silicate	30	4.507 μm	Paste	28	Room	Decreased porosity	15.4

(continued)

Table 3.1 (continued)

References	Precursor	Precursor size/fineness	Activator	MK (%)	MK size/fineness	Type	Age (day)	Curing	Effect	Ratio
	Slag	37.183 μm	NaOH and sodium silicate	30	4.507 μm	Paste	28	Water	Decreased porosity	18.1
Bernal et al. [12]	Slag	498 m²/kg	NaOH and sodium silicate solution	20, 40, 60 and 80	391 m²/kg	Mortar	28	25 °C with > 98% RH for 24 h, then at 27 °C with > 90% RH	Increased permeable pores volume	4.8, 44, 54.53 and 69.63
Alanazi et al. [13]	FA	–	NaOH and sodium silicate	10	–	Mortar	28	Room	Decreased porosity	10.69
Bheel et al. [14]	FA	379 m²/kg	NaOH and sodium silicate solution	20, 15, 10 and 5	18,000 m²/kg	Concrete	28	75 °C for 2 h	Decreased water penetration	60.95, 51.2, 33.33 and 19.04
Abbass et al. [15]	FA	37.92 μm	NaOH and sodium silicate	5 and 20	16.05 2.92	Paste	56	20 °C with 70% RH	Decreased porosity	–
	FA	37.92 μm	NaOH and sodium silicate	5 and 20	16.05 2.92	Paste	56	20 °C with 70% RH	Decreased water absorption	–
Zhang et al. [16]	FA	37 μm	NaOH and waterglass	10, 30, 50, 70 and 90	8 μm	Paste	28	40 °C for 48 h	Increased porosity	~1.8, ~2.1, ~2.8 and ~3.3 compared to those containing 10% MK

(continued)

Table 3.1 (continued)

References	Precursor	Precursor size/fineness	Activator	MK (%)	MK size/fineness	Type	Age (day)	Curing	Effect	Ratio
Zhang et al. [5]	FA	32 μm	KOH and potassium silicate	50 and 100	17 μm	Paste	7	20 °C with 90% RH	Decreased porosity	8.9 and 14.7
Gómez-Casero et al. [4]	FA	d50 19.87 μm	NaOH and sodium silicate	25, 50, 75 and 100	d50 9.579 μm	Paste	7	60 °C with standard humidity for one day	Increased porosity	~16, ~20.5, ~42.8 and 59.82
	FA	d50 19.87 μm	NaOH and sodium silicate	25, 50, 75 and 100	d50 9.579 μm	Paste	28	60 °C with standard humidity for one day	Decreased porosity	~0.8, 11.4, ~13.5 and 19.8
	FA	d50 19.87 μm	NaOH and sodium silicate	25, 50, 75 and 100	d50 9.579 μm	Paste	7	60 °C with standard humidity for one day	Increased water absorption	~25, ~42.58, ~57.14 and 67.14
	FA	d50 19.87 μm	NaOH and sodium silicate	25, 50, 75 and 100	d50 9.579 μm	Paste	28	60 °C with standard humidity for one day	Decreased water absorption	~0.7%, ~2.7%, ~7% and 7.4

(continued)

Table 3.1 (continued)

References	Precursor	Precursor size/fineness	Activator	MK (%)	MK size/ fineness	Type	Age (day)	Curing	Effect	Ratio
Bignozzi et al. [8]	Ladle slag	95% passing at 90 μm	NaOH and sodium silicate solution	20	95% passing at 30 μm	Paste	7	Room with > 90% RH, then at 20 ± 2 °C with 55% RH for 6 days	Decreased porosity	21.6
	Ladle slag	95% passing at 90 μm	NaOH and sodium silicate solution	30, 40, 50, 75 and 100	95% passing at 30 μm	Paste	7	Room with > 90% RH, then at 20 ± 2 °C with 55% RH for 6 days	Increased porosity	4.98, 5.65, 14.29, 28.24 and 41.86

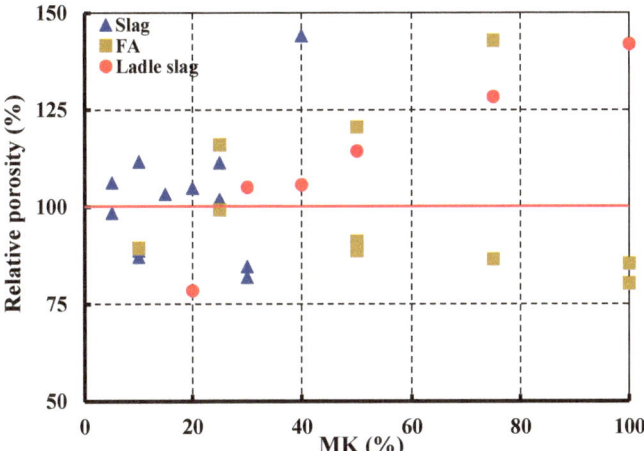

Fig. 3.1 Effect of MK on the porosity of geopolymers based on slag, FA and ladle slag precursors [1, 3–5, 8, 11–13]

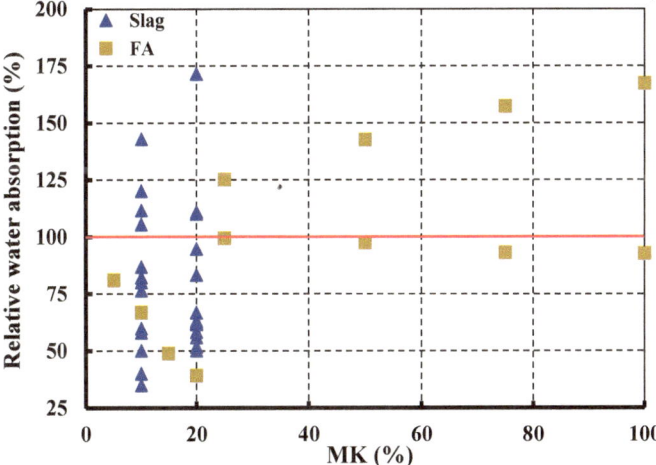

Fig. 3.2 Effect of MK on the water absorption of geopolymers based on slag and FA precursors [4, 9, 10, 14]

[11] found lower drying shrinkage of AAS mortars activated with NaOH and sodium silicate solution for up to 360 days by partially replacing slag with 5–25% MK. The drying shrinkage decreased with increasing MK content. After 360 days, the drying shrinkage of the control specimens was reduced by 14.25% with the inclusion of 25% MK.

Considering what has been reported thus far, it can be noticed that all available limited studies agreed that the incorporation of MK in the matrix decreased shrinkage.

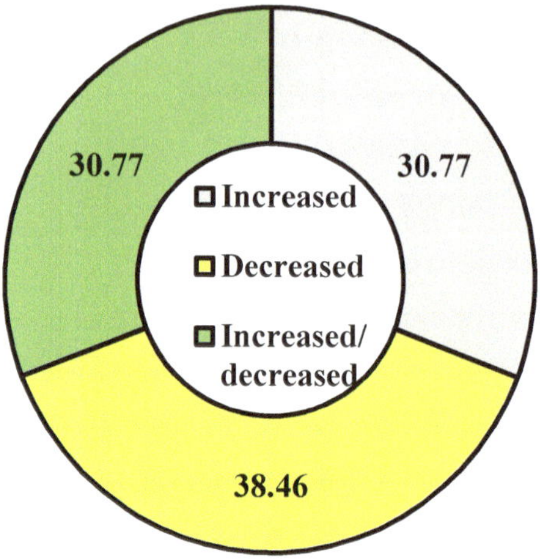

Fig. 3.3 Percentage of studies about the influence of MK on the porosity and water absorption of geopolymers based on slag, FA and ladle slag precursors

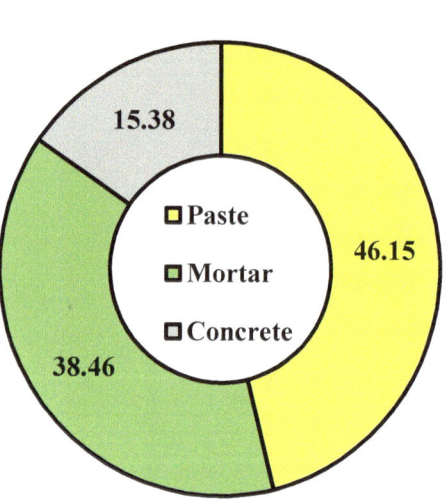

Fig. 3.4 Percentage of porosity and water absorption studies of various media of geopolymers based on different precursors containing MK

This reduction in the autogenous shrinkage could be attributed to the reduction in the self-desiccation of the matrix due to the increased porosity and the reduced chemical shrinkage with the incorporation of MK [18]. Li et al. [19] related this reduction to the reduction in the chemical shrinkage with including MK, which led to less air or gas occupying the volume of paste, and in the meantime, a coarser pore structure was obtained.

References

1. B. El Moustapha, S. Bonnet, A. Khelidj, N. Leklou, D. Froelich, I.A. Babah, C. Charbuillet, A. Khalifa, Compensation of the negative effects of micro-encapsulated phase change materials

by incorporating metakaolin in geopolymers based on blast furnace slag. Constr. Build. Mater. **314**, 125556 (2022)

2. G.F. Huseien, J. Mirza, M. Ismail, S. Ghoshal, M.A.M. Ariffin, Effect of metakaolin replaced granulated blast furnace slag on fresh and early strength properties of geopolymer mortar. Ain Shams Eng. J. **9**(4), 1557–1566 (2018)

3. M. Gharieb, Y.A. Mosleh, A.M. Rashad, Properties and corrosion behaviour of applicable binary and ternary geopolymer blends. Int. J. Sustain. Eng. **14**(5), 1068–1080 (2021)

4. M. Gómez-Casero, C. De Dios-Arana, J. Bueno-Rodríguez, L. Pérez-Villarejo, D. Eliche-Quesada, Physical, mechanical and thermal properties of metakaolin-fly ash geopolymers. Sustain. Chem. Pharm. **26**, 100620 (2022)

5. H.Y. Zhang, V. Kodur, B. Wu, L. Cao, S.L. Qi, Comparative thermal and mechanical performance of geopolymers derived from metakaolin and fly ash. J. Mater. Civ. Eng. **28**(2), 04015092 (2016)

6. A. Sharmin, U.J. Alengaram, M.Z. Jumaat, M.O. Yusuf, S.A. Kabir, I.I. Bashar, Influence of source materials and the role of oxide composition on the performance of ternary blended sustainable geopolymer mortar. Constr. Build. Mater. **144**, 608–623 (2017)

7. R. Rajamma, J.A. Labrincha, V.M. Ferreira, Alkali activation of biomass fly ash–metakaolin blends. Fuel **98**, 265–271 (2012)

8. M.C. Bignozzi, S. Manzi, I. Lancellotti, E. Kamseu, L. Barbieri, C. Leonelli, Mix-design and characterization of alkali activated materials based on metakaolin and ladle slag. Appl. Clay Sci. **73**, 78–85 (2013)

9. S.A. Bernal, R.M. De Gutiérrez, J.L. Provis, Engineering and durability properties of concretes based on alkali-activated granulated blast furnace slag/metakaolin blends. Constr. Build. Mater. **33**, 99–108 (2012)

10. B. El Moustapha, S. Bonnet, A. Khelidj, N. Maranzana, D. Froelich, A. Khalifa, I.A. Babah, Effects of microencapsulated phase change materials on chloride ion transport properties of geopolymers incorporating slag and, metakaolin, and cement-based mortars. J. Build. Eng., 106887 (2023)

11. M.A. Asaad, G.F. Huseien, R.P. Memon, S. Ghoshal, H. Mohammadhosseini, R. Alyousef, Enduring performance of alkali-activated mortars with metakaolin as granulated blast furnace slag replacement. Case Stud. Constr. Mater. **16**, e00845 (2022)

12. S. Bernal, R.M. de Gutiérrez, F. Ruiz, H. Quiñones, J. Provis, High-temperature performance of mortars and concretes based on alkali-activated slag/metakaolin blends. Mater. Constr. **62**(308), 471–488 (2012)

13. H. Alanazi, J. Hu, Y.-R. Kim, Effect of slag, silica fume, and metakaolin on properties and performance of alkali-activated fly ash cured at ambient temperature. Constr. Build. Mater. **197**, 747–756 (2019)

14. N. Bheel, P. Awoyera, T. Tafsirojjaman, N.H. Sor, Synergic effect of metakaolin and groundnut shell ash on the behavior of fly ash-based self-compacting geopolymer concrete. Constr. Build. Mater. **311**, 125327 (2021)

15. A.M. Abbass, R. Firdous, J.N. Yankwa Djobo, D. Stephan, M.A. Elrahman, The role of chemistry and fineness of metakaolin on the fresh properties and heat resistance of blended fly ash-based geopolymer. SN Appl. Sci. **5**(5), 1–17 (2023)

16. X. Zhang, X. Zhang, X. Li, M. Ma, Z. Zhang, X. Ji, Slurry rheological behaviors and effects on the pore evolution of fly ash/metakaolin-based geopolymer foams in chemical foaming system with high foam content. Constr. Build. Mater. **379**, 131259 (2023)

17. A.M. Rashad, S.R. Zeedan, Effect of silica fume and activator concentration on metakaolin geopolymer exposed to thermal loads. ACI Mater. J. 120(1) (2023)

18. Z. Li, X. Liang, Y. Chen, G. Ye, Effect of metakaolin on the autogenous shrinkage of alkali-activated slag-fly ash paste. Constr. Build. Mater. **278**, 122397 (2021)

19. Z. Li, M. Nedeljković, B. Chen, G. Ye, Mitigating the autogenous shrinkage of alkali-activated slag by metakaolin. Cem. Concr. Res.Concr. Res. **122**, 30–41 (2019)

Chapter 4
Effect of Metakaolin on the Mechanical Properties of Geopolymers

4.1 Compressive Strength

4.1.1 Slag Precursor

Huseien et al. [1] found 15.85%, 26.22% and 37.8% reduction in the 1 day compressive strength of alkali-activated slag (AAS) mortars activated with NaOH and sodium silicate solution by partially replacing slag with 5%, 10% and 15% MK, respectively, whilst the 7 days compressive strength was enhanced by ~57.14%, ~57.2% and ~64.28%, respectively. The incorporation of 5%, 10% and 15% MK increased the 28 days compressive strength by 39.5%, 40.18% and 40.85%, respectively. Rashad [2] found 24.57% reduction in the 28 days compressive strength of slag pastes activated with Na_2SO_4 by partially replacing slag (fineness 450 m^2/kg) with 10% MK (fineness 14,700 m^2/kg). The specimens were cured at 40 °C. Li et al. [3] partially replaced slag (size 0.1–50 μm) with 10% and 20% MK (size 0.15–142 μm) in pastes activated by NaOH and sodium silicate solution. The specimens were cured at 20 °C. The results showed a general reduction in compressive strength with the incorporation of MK. The incorporation of 10% MK decreased the 1, 7 and 28 days compressive strength by ~11.9%, ~2.4% and ~7.2%, respectively, whilst the incorporation of 20% MK decreased it by ~26.8%, ~23% and ~15.5%, respectively. Bernal et al. [4] prepared AAS mortars activated with sodium silicate. The slag (size 0.1–74 μm) was partially replaced with 10% and 20% MK (size 1.8–100 μm). The specimens were cured at 27 ± 2 °C with 90% RH. The results showed a reduction in compressive strength up to 180 days with the inclusion of MK. This reduction increased when Si/Na increased from 2 to 2.4. When Si/Na was 2, the incorporation of 10% MK decreased the 28, 60 and 180 days by ~5.7%, ~5.6% and ~1.3%, respectively, whilst the incorporation of 20% MK decreased the 28, 60 and 180 days by ~15.7%, ~14.1% and ~12%, respectively. When the Si/Na was 2.4, The incorporation of 10% MK decreased the 28,

60 and 180 days by ~11.7%, ~13.85% and ~15.7%, respectively, whilst the incorporation of 20% MK decreased the 28, 60 and 180 days by ~23.3%, ~13.85% and
~15.7%, respectively.

Bernal et al. [5] prepared AAS mortars activated with NaOH and sodium silicate
solution. The slag (size 0.1–74 μm d50 15 μm) was partially replaced with 10%
and 20% MK (size 1.8–100 μm d50 12.2 μm). Different Si/Na ratios were used.
The specimens were cured at 27 ± 2 °C with 90% RH. When the ratio of Si/Na was
1.6, the 28 days compressive strength was enhanced by ~48.1% with the inclusion
of 10% MK, whilst the inclusion of 20% MK decreased it by ~62.9%. When the
ratio of Si/Na was 4, the 28 days compressive strength was decreased by ~20.5% and
~69.1% with the inclusion of 10% and 20% MK, respectively. When the ratio of Si/
Na was 4.4, the 28 days compressive strength was decreased by ~8.8% and ~42.6%
with the inclusion of 10% and 20% MK, respectively. Bernal et al. [6] prepared AAS
concrete specimens activated with NaOH and sodium silicate solution. The slag (size
0.1–74 μm d50 15 μm) was partially replaced with 10% and 20% MK (size 1.8–
100 μm d50 12.2 μm). Different ratios of Si/Al were used. The specimens were
cured at 25 ± 5 °C with 90% RH. When the Si/Al ratio was 3.6, the incorporation of
10% MK increased the 7, 28, 90 and 180 days compressive strength by ~141.17%,
~127.27%, ~87.1% and ~73.53%, respectively, whilst the incorporation of 20% MK
increased it by ~164%, ~131,8%, ~187% and ~85.3%, respectively. When the Si/
Al ratio was 4, the incorporation of 10% MK increased the 7, 28, 90 and 180 days
compressive strength by ~101.6%, ~104%, ~67.7% and ~64.6%, respectively, whilst
the incorporation of 20% MK increased it by ~53.3%, ~72.8%, ~49.5% and ~52.1%,
respectively. When the Si/Al ratio was 4.4, the incorporation of 10% MK increased
the 7, 28 and 180 days by ~46.25%, ~14.75% and ~6.75%, respectively, whilst the 90
days compressive strength was decreased by ~7.2%. The incorporation of 20% MK
decreased the 7 and 28 days compressive strength by ~0% and ~3.3%, respectively,
whilst the 90 and 180 days compressive strength was enhanced by ~0% and ~6.75%,
respectively. Bernal [7] prepared AAS mortars activated with NaOH and sodium
silicate solution. Different SiO_2/Al_2O_3 ratios of 3.6, 4 and 4.4 were used. The slag
(fineness 399 m^2/kg) was partially replaced with 10% and 20% MK (fineness 391
m^2/kg). The specimens were cured at 27 ± 2 °C with 90% RH. The incorporation of
10% MK increased the 7 days compressive strength by ~116%, ~16.67% and 44%
when the SiO_2/Al_2O_3 ratio was 3.6, 4 and 4.4, respectively, whilst the incorporation
of 20% MK enhanced it by ~196%, ~38.7% and ~44%, respectively.

El Moustapha et al. [8] found 12.28% and 16.1% enhancement in the 28 and 90
days compressive strength of AAS mortars activated with NaOH and sodium silicate
solution by partially replacing slag with 20% MK, whilst the incorporation of 10%
MK did not show a fundamental change in the strength. The specimens were cured
at room temperature. Chen and Ye [9] found ~64.7%, ~30.43%, ~21.43%, 36.36%
and ~38.9% enhancement in the 1, 3, 7, 14 and 28 days compressive strength of
AAS pastes activated with NaOH and MgO by partially replacing slag (fineness 501
m^2/kg) with 20% MK, respectively. The specimens were sealed and cured at 23.0
± 2.0 °C with 90% RH. Asaad et al. [10] found 15.8% enhancement in the 1 day

compressive strength of AAS mortars activated with NaOH and sodium silicate solution by partially replacing slag (size 12 μm) with 5% MK (size 10 μm), whilst the incorporation of 10% and 25% MK decreased it by 6.27% and 32.23%, respectively. At the age of 7 days, the incorporation of 5% MK increased the compressive strength by 2.53%, whilst the incorporation of 25% decreased it by 18.95%. At the age of 28 days, the incorporation of 5% and 10% MK increased the compressive strength by 2.8% and 3.6%, respectively. At the age of 360 days, the incorporation of 5%, 10%, 15% and 20% MK increased the compressive strength by 10.64%, 12.92%, 6.7% and 2.13%, respectively, whilst the incorporation of 25% MK decreased it by 0.61%. All specimens were cured at 27 °C with 75% RH. Gharieb et al. [11] found 33.1%, 29%, 44.83% and 41% enhancement in the 1, 3, 7 and 28 days compressive strength of AAS pastes activated with NaOH and sodium silicate solution by partially replacing slag (average diameter = 37.183μm) with 30% MK (average diameter = 4.507 μm), respectively, when air curing was used. When water curing was used, the enhancement in the strength was 33.1%, 24.36%, 8.15% and 7.75%, respectively. Kumar et al. [12] found 9.6%, 17.48% and 2% enhancement in the 7 days compressive strength of AAS concretes activated with NaOH and sodium silicate solution by partially replacing slag (fineness 370 m^2/kg) with 10%, 20% and 30% MK, respectively, whilst the 28 days compressive strength was enhanced by 13%, 24.36% and 2.87%, respectively. On the other hand, the incorporation of 40% MK decreased the 7 and 28 days compressive strength by 24.89% and 16.94%, respectively. Khalil et al. [13] found a general reduction in the compressive strength of AAS mortars activated with NaOH and sodium silicate solution by partially replacing slag with 50% MK. The specimens were cured at room temperature. When the SiO_2/Na_2O ratio was 1.1, 1.3 and 1.7, the reduction in the 7 days compressive strength was 26.4%, 38% and 12.24%, respectively, whilst the reduction in the 28 days compressive strength was 19.29%, 44.74% and 38.62%, respectively, whilst the reduction in the 90 days compressive strength was 17.79%, 50.29% and 39.55%, respectively. Bernal et al. [14] found 15.18% and 6.12% reduction in the 7 and 28 days compressive strength of AAS concretes by partially replacing slag (fineness 498 m^2/kg) with 60% MK (fineness 391 m^2/kg), respectively. They also found a general reduction in the 7, 28 and 60 days compressive strength of AAS mortars by partially replacing slag with 20–80% MK.

Burciaga-Díaz et al. [15] prepared AAS pastes activated with NaOH and sodium silicate solution. The slag (fineness 465.3 m^2/kg) was replaced by MK (fineness 946.9 m^2/kg). The specimens were cured at 20 °C. When the activator concentration was 5% of Na_2O, the incorporation of 20%, 50%, 80% and 100% MK decreased the compressive strength by ~41.2%, ~77.2%, 86.7% and 100%, respectively. When the activator concentration was 10% of Na_2O, the incorporation of 20% and 50% MK increased the compressive strength by ~50% and ~22.2%, whilst the incorporation of 80% and 100% MK decreased it by ~28.9% and ~80%, respectively. When the activator concentration was 15% of Na_2O, the incorporation of 20%, 50%, 80% and 100% MK increased the compressive strength by ~27.6%, ~210%, 151.7% and 231%, respectively. Burciaga-Díaz et al. [16] replaced slag (fineness 465.3 m^2/kg) by 20–100% MK (fineness 946.9 m^2/kg) in pastes activated with NaOH and sodium silicate solution. The results showed ~8.57%, 35.71%, ~62.85% and 71.43% reduction in the compressive strength with the incorporation of 20%, 50%, 80% and 100% MK, respectively. Peng et al. [17] prepared AAS pastes activated with NaOH and sodium silicate solution. The slag (BET fineness 15 m^2/g) was replaced with 20–100% MK (BET fineness 20 m^2/g). Different Si/Na ratios of 1.4–2 were used. The specimens were cured at ambient temperature. When the Si/Na ratio was 1.4, the 7 days compressive strength was decreased by 22.5%, 7.86%, 27.7% and 88.15% with the inclusion of 20%, 60%, 80% and 100% MK, respectively, whilst the incorporation of 40% MK increased it by 18.7%. The incorporation of 20% and 100% MK decreased the 28 days compressive strength by 4.78% and 78.68%, respectively, whilst the incorporation of 40%, 60% and 80% MK increased it by 40.58%, 72.98% and 53.35%, respectively. When the Si/Na ratio of 1.6 was used, the 7 days compressive strength was increased by 11.32%, 49.5% and 16.7% with the incorporation of 20%, 40% and 60% MK, respectively, whilst the incorporation of 80% and 100% decreased it by 23.1% and 90.1%, respectively. The 28 days compressive strength was decreased by 1.34%, 4.96% and 87.68% with the inclusion of 20%, 80% and 100% MK, respectively, whilst the incorporation of 40% and 60% MK increased it by 3.7% and 31.74%, respectively. When the Si/Na ratio of 1.8 was used, the incorporation of 20% and 40% MK increased the 7 days compressive strength by 16.16% and 37.24%, respectively, whilst the incorporation of 60%, 80% and 100% MK decreased it by 1.7%, 66.5% and 93.5%, respectively. The 28 days compressive strength was decreased by 2.22%, 8.18%, 4.57%, 73.81% and 92.63% with the incorporation of 20%, 40%, 60%, 80% and 100% MK, respectively. When the Si/Na ratio of 2 was used, the incorporation of 20%, 60%, 80% and 100% MK decreased the 7 days compressive strength by 2.48%, 49.97%, 79.17% and 97.59%, respectively, whilst the incorporation of 40% increased it by 16.2%. The incorporation of 20%, 40%, 60%, 80% and 100% MK decreased the 28 days compressive strength by 7.9%, 3.82%, 31.14%, 76.1% and 97.59%, respectively. Buchwald et al. [18] prepared AAS pastes activated with NaOH. The slag (size 15.8 μm) was replaced with 25–100% MK (size 3.9 μm). The specimens were cured at 40 °C. The 28 days compressive strength was decreased by ~17.14%, ~11.4% and ~85.7% with the incorporation of 25%, 50% and 100% MK, respectively. Burciaga-Díaz et al. [19] found 36.96% and 55.43% reduction in the 3 days compressive strength of AAS pastes activated with NaOH and sodium silicate solution by replacing slag (fineness 465.3 m^2/g) with 50% and 100% MK (fineness 946.9 m^2/g), respectively. The specimens were cured at 20 °C for 24 h with 80% RH, then at 60 °C for 48 h. The results included in this section are summarized in Table 4.1.

Considering what has been reported thus far, it is clear that the findings on the effect of MK on AAS matrices compressive strength are inconclusive (Fig. 4.1), but most of them (around 45%) confirmed lower compressive strength with the incorporation of MK [2–4, 13, 14, 16, 18, 19], whilst around 20% of them confirmed higher compressive strength with the incorporation of MK [7–9, 11]. The remaining studies (around 35%) reported that the incorporation of MK may decrease or increase the compressive strength [1, 5, 6, 10, 12, 15, 17] as shown in Fig. 4.2. This typically depends on testing age, activator type/concentration, Si/Na ratio, Si/Al ratio, MK

Table 4.1 Effect of MK on the compressive strength of AAS matrices

References	Precursor	Precursor size/fineness	Activator	MK (%)	MK size/fineness	Type	Age (day)	Effect	Ratio
Huseien et al. [1]	Slag	60% < 10 μm	NaOH and sodium silicate solution	5, 10 and 15	75% < 10 μm	Mortar	1	Decreased	15.85, 26.22 and 37.8
	Slag	60% < 10 μm	NaOH and sodium silicate solution	5, 10 and 15	75% < 10 μm	Mortar	7	Increased	~57.14, ~57.2 and ~64.28
	Slag	60% < 10 μm	NaOH and sodium silicate solution	5, 10 and 15	75% < 10 μm	Mortar	28	Increased	39.5, 40.18 and 40.85
Rashad [2]	Slag	450 m²/kg	Na₂SO₄	10	14,700 m²/kg	Paste	28	Decreased	24.57
Li et al. [3]	Slag	0.1–50 μm	NaOH and sodium silicate solution	10	0.15–142 μm	Paste	1, 7 and 28	Decreased	~11.9, ~2.4 and 7.2
	Slag	0.1–50 μm	NaOH and sodium silicate solution	20	0.15–142 μm	Paste	1, 7 and 28	Decreased	~26.8, ~23 and ~15.5
Bernal et al. [4]	Slag	0.1–74 μm	Sodium silicate	10	1.8–100 μm	Mortar	28, 60 and 180	Decreased	~5.7, ~5.6 and ~1.3 (Si/Na = 2)
	Slag	0.1–74 μm	Sodium silicate	20	1.8–100 μm	Mortar	28, 60 and 180	Decreased	~15.7, ~14.1 and ~12 (Si/Na = 2)

(continued)

Table 4.1 (continued)

References	Precursor	Precursor size/fineness	Activator	MK (%)	MK size/fineness	Type	Age (day)	Effect	Ratio
	Slag	0.1–74 μm	Sodium silicate	10	1.8–100 μm	Mortar	28, 60 and 180	Decreased	~11.7, ~13.85 and ~15.7 (Si/Na = 2.4)
	Slag	0.1–74 μm	Sodium silicate	20	1.8–100 μm	Mortar	28, 60 and 180	Decreased	~23.3, ~13.85 and ~15.7 (Si/Na = 2.4)
Bernal et al. [5]	Slag	0.1–74 μm	NaOH and sodium silicate	10	1.8–100 μm	Mortar	28	Increased	~48.1 (Si/Na = 1.6)
	Slag	0.1–74 μm	NaOH and sodium silicate	20	1.8–100 μm	Mortar	28	Decreased	~62.9 (Si/Na = 1.6)
	Slag	0.1–74 μm	NaOH and sodium silicate	10 and 20	1.8–100 μm	Mortar	28	Decreased	~20.5 and ~69.1 (Si/Na = 4)
	Slag	0.1–74 μm	NaOH and sodium silicate	10 and 20	1.8–100 μm	Mortar	28	Decreased	~8.8 and ~42.6 (Si/Na = 4.4)
Bernal et al. [6]	Slag	0.1–74 μm	NaOH and sodium silicate	10	1.8–100 μm	Concrete	7, 28, 90 and 180	Increased	~141.17, ~127.27, ~87.1 and ~73.53 (Si/Al = 3.6)
	Slag	0.1–74 μm	NaOH and sodium silicate	20	1.8–100 μm	Concrete	7, 28, 90 and 180	Increased	~164, ~131.8, ~187 and ~85.3 (Si/Al = 3.6)

(continued)

Table 4.1 (continued)

References	Precursor	Precursor size/fineness	Activator	MK (%)	MK size/fineness	Type	Age (day)	Effect	Ratio
	Slag	0.1–74 μm	NaOH and sodium silicate	10	1.8–100 μm	Concrete	7, 28, 90 and 180	Increased	By ~101.6, ~104, ~67.7 and ~64.6 (Si/Al = 4)
	Slag	0.1–74 μm	NaOH and sodium silicate	20	1.8–100 μm	Concrete	7, 28, 90 and 180	Increased	~53.3, ~72.8%, ~49.5% and ~52.1 (Si/Al = 4)
	Slag	0.1–74 μm	NaOH and sodium silicate	10	1.8–100 μm	Concrete	7, 28 and 180	Increased	~46.25, ~14.75 and ~6.75 (Si/Al = 4.4)
	Slag	0.1–74 μm	NaOH and sodium silicate	10	1.8–100 μm	Concrete	90	Decreased	~7.2 (Si/Al = 4.4)
	Slag	0.1–74 μm	NaOH and sodium silicate	20	1.8–100 μm	Concrete	7 and 28	Decreased	~0 and ~3.3 (Si/Al = 4.4)
	Slag	0.1–74 μm	NaOH and sodium silicate	20	1.8–100 μm	Concrete	90 and 180	Increased	0% and ~6.75 (Si/Al = 4.4)
Bernal [7]	Slag	399 m²/kg	NaOH and sodium silicate	10	391 m²/kg	Mortar	7	Increased	~116, ~16.67 and 44 (when SiO_2/Al_2O_3 = 3.6, 4 and 4.4, respectively)
	Slag	399 m²/kg	NaOH and sodium silicate	20	391 m²/kg	Mortar	7	Increased	~196, ~38.7 and ~44 (when SiO_2/Al_2O_3 = 3.6, 4 and 4.4, respectively)

(continued)

Table 4.1 (continued)

References	Precursor	Precursor size/fineness	Activator	MK (%)	MK size/fineness	Type	Age (day)	Effect	Ratio
El Moustapha et al. [8]	Slag	445 m²/kg	NaOH and sodium silicate	10	17,000 m²/kg	Mortar	28 and 90	No effect	–
	Slag	445 m²/kg	NaOH and sodium silicate	20	17,000 m²/kg	Mortar	28 and 90	Increased	12.28 and 16.1
Chen and Ye [9]	Slag	501 m²/kg	NaOH and MgO	20	> 501 m²/kg	Paste	1, 3, 7, 14 and 28	Increased	64.7, ~30.43, ~21.43, 36.36 and ~38.9
Asaad et al. [10]	Slag	12 μm	NaOH and sodium silicate	5	12 μm	Mortar	1	Increased	15.8
	Slag	12 μm	NaOH and sodium silicate	10 and 25	12 μm	Mortar	1	Decreased	6.27 and 32.23
Gharieb et al. [11]	Slag	37.183μm	NaOH and sodium silicate	30	4.507 μm	Paste	1, 3, 7 and 28	Increased	33.1, 29, 44.83 and 41
Kumar et al. [12]	Slag	370 m²/kg	NaOH and sodium silicate	10, 20 and 30	–	Concrete	7	Increased	9.6, 17.48 and 2
	Slag	370 m²/kg	NaOH and sodium silicate	10, 20 and 30	–	Concrete	28	Increased	13, 24.36 and 2.87
	Slag	370 m²/kg	NaOH and sodium silicate	40	–	Concrete	7 and 28	Decreased	24.89 and 16.94

(continued)

Table 4.1 (continued)

References	Precursor	Precursor size/fineness	Activator	MK (%)	MK size/fineness	Type	Age (day)	Effect	Ratio
Khalil et al. [13]	Slag	408.8 m^2/kg	NaOH and sodium silicate solution	50	18,000 m^2/kg	Mortar	7	Decreased	26.4, 38 and 12.24 (SiO_2/Na_2O = 1.1, 1.3 and 1.7, respectively)
	Slag	408.8 m^2/kg	NaOH and sodium silicate solution	50	18,000 m^2/kg	Mortar	28	Decreased	19.29, 44.74 and 38.62 (SiO_2/Na_2O = 1.1, 1.3 and 1.7, respectively)
	Slag	408.8 m^2/kg	NaOH and sodium silicate solution	50	18,000 m^2/kg	Mortar	90	Decreased	17.79, 50.29 and 39.55 (SiO_2/Na_2O = 1.1, 1.3 and 1.7, respectively)
Bernal et al. [14]	Slag	498 m^2/kg	NaOH and sodium silicate solution	60	391 m^2/kg	Concrete	7 and 28	Decreased	15.18 and 6.12
	Slag	498 m^2/kg	NaOH and sodium silicate solution	20, 40, 60 and 80	391 m^2/kg	Mortar	7	Decreased	~0, ~26.3, ~21.1 and ~14.5
	Slag	498 m^2/kg	NaOH and sodium silicate solution	20, 40, 60 and 80	391 m^2/kg	Mortar	28	Decreased	~0, ~35, ~14.5 and ~31.6
	Slag	498 m^2/kg	NaOH and sodium silicate solution	20, 40, 60 and 80	391 m^2/kg	Mortar	60	Decreased	~6, ~6, ~19.3 and ~31.9
Burciaga-Díaz et al. [15]	Slag	465.3 m^2/kg	NaOH and sodium silicate	20, 50, 80 and 100	946.9 m^2/kg	Paste	28	Decreased	~41.2, ~77.2, 86.7% and 100 (Na_2O = 5)

(continued)

Table 4.1 (continued)

References	Precursor	Precursor size/fineness	Activator	MK (%)	MK size/fineness	Type	Age (day)	Effect	Ratio
	Slag	465.3 m²/kg	NaOH and sodium silicate	20 and 50	946.9 m²/kg	Paste	28	Increased	~50 and ~22.2 ($Na_2O = 10$)
	Slag	465.3 m²/kg	NaOH and sodium silicate	80 and 100	946.9 m²/kg	Paste	28	Decreased	~28.9 and ~80 ($Na_2O = 10$)
	Slag	465.3 m²/kg	NaOH and sodium silicate	20, 50, 80 and 100	946.9 m²/kg	Paste	28	Increased	~27.6, ~210, 151.7 and 231 ($Na_2O = 15$)
Burciaga-Díaz et al. [16]	Slag	465.3 m²/kg	NaOH and sodium silicate	20, 50, 80 and 100	946.9 m²/kg	Paste	28	Decreased	~8.57, 35.71, ~62.85 and 71.43
Peng et al. [17]	Slag	BET 15 m²/g	NaOH and sodium silicate	20, 60, 80 and 100	BET 20 m²/g	Paste	7	Decreased	22., 7.86, 27.7 and 88.15 (Si/Na = 1.4)
	Slag	BET 15 m²/g	NaOH and sodium silicate	40	BET 20 m²/g	Paste	7	Increased	18.7 (Si/Na = 1.4)
	Slag	BET 15 m²/g	NaOH and sodium silicate	20 and 100	BET 20 m²/g	Paste	28	Decreased	4.78 and 78.68 (Si/Na = 1.4)
	Slag	BET 15 m²/g	NaOH and sodium silicate	40, 60 and 80	BET 20 m²/g	Paste	28	Increased	40.58, 72.98 and 53.35 (Si/Na = 1.4)
	Slag	BET 15 m²/g	NaOH and sodium silicate	20, 40 and 60	BET 20 m²/g	Paste	7	Increased	11.32, 49.5 and 16.7 (Si/Na = 1.6)
	Slag	BET 15 m²/g	NaOH and sodium silicate	80 and 100	BET 20 m²/g	Paste	7	Decreased	23.1 and 90.1 (Si/Na = 1.6)

(continued)

Table 4.1 (continued)

References	Precursor	Precursor size/fineness	Activator	MK (%)	MK size/fineness	Type	Age (day)	Effect	Ratio
	Slag	BET 15 m²/g	NaOH and sodium silicate	20, 80 and 100	BET 20 m²/g	Paste	28	Decreased	1.34, 4.96 and 87.68 (Si/Na = 1.6)
	Slag	BET 15 m²/g	NaOH and sodium silicate	40 and 60	BET 20 m²/g	Paste	28	Increased	3.7% and 31.74 (Si/Na = 1.6)
	Slag	BET 15 m²/g	NaOH and sodium silicate	20 and 40	BET 20 m²/g	Paste	7	Increased	16.16 and 37.24 (Si/Na = 1.8)
	Slag	BET 15 m²/g	NaOH and sodium silicate	60, 80 and 100	BET 20 m²/g	Paste	7	Decreased	1.7, 66.5 and 93.5 (Si/Na = 1.8)
	Slag	BET 15 m²/g	NaOH and sodium silicate	20, 40, 60, 80 and 100	BET 20 m²/g	Paste	28	Decreased	2.22, 8.18, 4.57, 73.81 and 92.63 (Si/Na = 1.8)
	Slag	BET 15 m²/g	NaOH and sodium silicate	20, 60, 80 and 100	BET 20 m²/g	Paste	7	Decreased	2.48, 49.97, 79.17 and 97.59 (Si/Na = 2)
	Slag	BET 15 m²/g	NaOH and sodium silicate	40	BET 20 m²/g	Paste	7	Increased	16.2
	Slag	BET 15 m²/g	NaOH and sodium silicate	20, 40, 60, 80 and 100	BET 20 m²/g	Paste	28	Decreased	7.9, 3.82, 31.14, 76.1 and 97.59
Buchwald et al. [18]	Slag	15.8 µm	NaOH	25, 50 and 100	3.9 µm	Paste	28	Decreased	~17.14, ~11.4 and ~85.7
Burciaga-Díaz et al. [19]	Slag	465.3 m²/g	NaOH and sodium silicate	50 and 100	946.9 m²/g	Paste	3	Decreased	36.96 and 55.43

ratio/fineness, slag fineness and curing condition. Broadly speaking, it is better to not increase the ratio of MK by more than 30% to obtain acceptable compressive strength, but 20% seems to be better than 30% as shown in Fig. 4.1. Huseien et al. [1] related this improvement to the increase in Al_2O_3 and SiO_2 content that led to improving geopolymerization and the formation of N–A–S–H as well as C–A–S–H. Asaad et al. [10] related this enhancement to the formation of denser gels, condensation reaction hastening with the incorporation of MK. Because MK has high ratios of SiO_2 and Al_2O_3, enhanced geopolymerization reaction was obtained. This can produce more N–A–S–H and C–A–S–H gels in addition to C–S–H. However, higher ratios of MK than 30% may cause adverse effect on the strength. This reduction could be attributed to the incomplete reaction of MK in the system due to insufficient activation conditions [5] or the coarser pore structure and lower amounts of reaction product formation with the inclusion of MK [3]. In actuality, the mentioned references in this section that focused on concrete, paste and mortar are roughly 15%, 40% and 45%, respectively (Fig. 4.3).

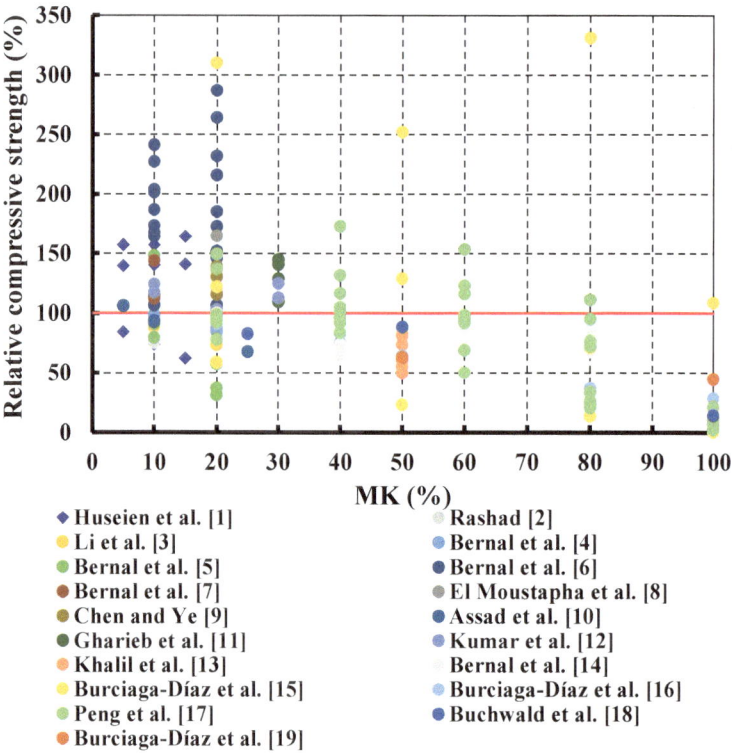

Fig. 4.1 Effect of MK on the compressive strength of slag geopolymers [1–19]

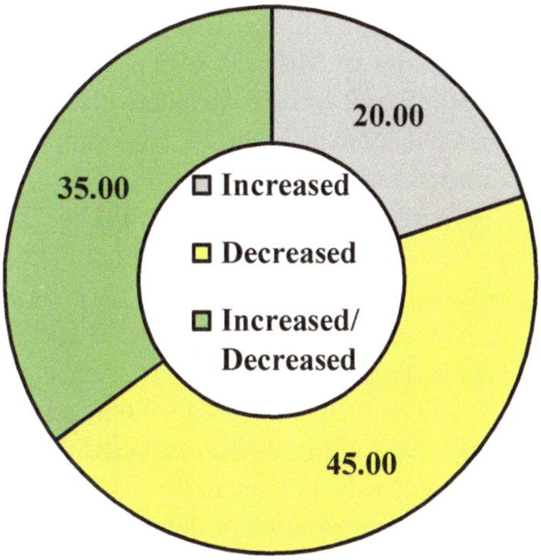

Fig. 4.2 Percentage of studies about the influence of MK on the compressive strength of slag geopolymers

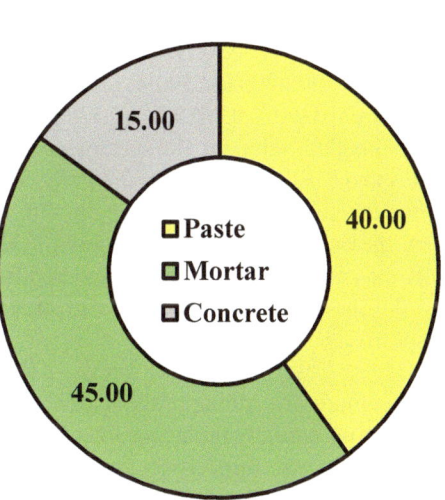

Fig. 4.3 Percentage of compressive strength studies of various media based on slag geopolymers containing MK

4.1.2 Fly Ash Precursor

Alanazi et al. [20] found ~30%, ~42.8%, ~45% and ~0% enhancement in the 1, 3, 7 and 28 days compressive strength of FA mortars activated with NaOH and sodium silicate solution by partially replacing FA with 5% MK, respectively, whilst including 10% MK increased it by ~60%, ~100%, ~32.5% and ~6.75%, respectively, when the sodium silicate/NaOH ratio was 1. When the sodium silicate/NaOH ratio was 2.5, the incorporation of 5% MK increased the 1, 3, 7 and 28 days compressive strength by ~87.5%, ~56.67%, 18% and ~0%, respectively, whilst including 10% MK increased it by ~62.5%, ~106.6%, ~46% and 1.28%, respectively. Abbass et al. [21] found ~163.6% and ~472.7% enhancement in the 56 days compressive strength of FA geopolymer pastes activated with NaOH and sodium silicate solution by partially replacing FA (size 37.92 μm) with 5% and 20% MK (size 16.05 μm), respectively. When the size of MK was 2.92 μm), the enhancement in the compressive strength was ~500% and ~518%, respectively. Bheel et al. [22] found ~5.35%, ~8.9% and ~2.3% enhancement in the 28 days compressive strength of self-compacting concrete specimens activated with NaOH and sodium silicate solution cured at 75 °C for 2 h by partially replacing FA (fineness 379 m²/kg) with 5%, 10% and 15% MK (fineness 18,000 m²/kg). On the other hand, the incorporation of 20% MK decreased the strength by ~6.25%. Duan et al. [23] partially replaced FA (d50 51.09 μm) by 5–20 MK (d50 12.66 μm) in paste specimens activated with NaOH and sodium silicate solution. The specimens were cured at 20 ± 2 °C with 90 ± 5% RH for 28 days. The results showed ~9.6%, ~16.5%, ~28.7% and ~39.13% enhancement in the 28 days compressive strength with the inclusion of 5%, 10%, 15% and 20% MK, respectively. Guo et al. [24] prepared FA/MK pastes activated with phosphoric acid. The specimens were cured at 40 °C with 90% RH for six days and then at 80 °C for 24 h. The results showed 292%, 184.9% and 246.5% enhancement in the 7 days compressive strength of FA pastes by partially replacing FA with 20%, 30% and 50% MK, respectively, whilst the 100 days compressive strength was enhanced by 285.7%, 307.9% and 590%, respectively. Zhang et al. [25] prepared geopolymer pastes from 0.9/0.1, 0.7/0.3, 0.5/0.5, 0.3/0.7 and 0.1/0.9 FA/MK activated with NaOH and waterglass solution. The specimens were cured at 40 °C for 28 h. The grain sizes of FA and MK were 37 μm and 8 μm, respectively. The results showed an increase in the 28 days compressive strength with increasing MK ratios from 10% to 50%, then the compressive strength started to decrease with including higher ratios of MK than 50%. The specimens prepared from 0.9/0.1, 0.7/0.3, 0.5/0.5, 0.3/0.7 and 0.1/0.9 showed 28 days compressive strength of 4.7, ~20, 25.2, ~22 and ~17 MPa, respectively.

Gómez-Casero et al. [26] found ~22.2%, ~8.3% and ~2.7% reduction in the 7 days compressive strength of FA pastes activated with NaOH and sodium silicate solution by partially replacing FA (d50 19.87 μm) with 25%, 50% and 75% MK (d50 9.579 μm), respectively, whilst replacing FA with 100% MK increased the strength by 16.67%. The 28 days compressive strength was decreased by ~31.2%, ~20% and ~15.3% with the incorporation of 25%, 50% and 75% MK, respectively, whilst the incorporation of 100% MK increased it by 16.4%. The specimens were cured at 60 °C with standard humidity for one day. Fu et al. [27] activated FA paste specimens with NaOH or KOH. The FA (size 19.3 μm) was replaced with 20–100% MK (size 1.8 μm). The specimens were cured in sealed bags at 50 °C for 24 h, then subjected to standard curing (20 ± 0.5 °C with 100% RH). When NaOH was used, the incorporation of 20% MK decreased the 1, 3, 7 and 28 days compressive strength by ~70.6%, ~73.8%, 75% and 77.3%, respectively, whilst the incorporation of 50% MK decreased it by ~85.3%, ~83.3%, ~84% and ~85%, respectively, and the incorporation of 100% MK decreased it by ~89.7%, ~88.1%, ~88.6% and 92.4%, respectively. When KOH was used, the incorporation of 20% MK decreased the 1, 3 and 7 days compressive strength by ~11.11%, ~8% and 0%, respectively, whilst the 28 days compressive strength was increased by ~5.7%. The incorporation of 50% MK decreased the 1, 3, 7 and 28 days compressive strength by 44.4%, ~52%,

~48.4% and ~62.3%, respectively, whilst the incorporation of 100% MK decreased it by ~38.9%, ~44%, ~45% and 66%, respectively. Zhang et al. [28] prepared FA geopolymer pastes activated with KOH and potassium silicate solution. The FA (size 32 μm) was replaced by 25–100% MK (size 17 μm). The specimens were cured at 20 °C with 90% RH for 7 days. The results showed ~97.4%, ~250%, ~229% and ~229% enhancement in the compressive strength with the inclusion of 25%, 50%, 80% and 100% MK, respectively. The results included in this part are summarized in Table 4.2.

Considering what has been reported thus far, it is clear that the findings on the effect of MK on FA geopolymer compressive strength are inconclusive (Fig. 4.4), but most of them (around 66.67%) confirmed higher compressive strength with the incorporation of MK [20, 21, 23, 24, 28], whilst the remaining (around 33.33%) reported that the incorporation of MK may decrease or increase the compressive strength [22, 26, 27], as shown in Fig. 4.5. This typically depends on testing age, activator type/concentration, Si/Al ratio, Na/Al ratio, MK ratio/fineness, slag fineness and curing condition. Broadly speaking, it is better to not increase the ratio of MK by more than 30% to obtain better compressive strength as shown in Fig. 4.4. Bheel et al. [22] related this enhancement to the fine MK particles that can refine the pore size, increase packing and produce a denser matrix. Guo et al. [24] related this enhancement to the high aluminium ratio in MK. Zhang et al. [28] related this enhancement to the higher polymerization reaction with including MK. However, higher ratios of MK than 30% may decrease the compressive strength due to the formation of a less compacted microstructure [27]. In actuality, the mentioned references in this section that focused on paste, concrete and mortar are roughly 77.78%, 11.11% and 11.11%, respectively (Fig. 4.6).

4.1.3 Other Precursors

Pascual et al. [29] prepared waste glass powder geopolymer mortars activated with NaOH. The powder (size 10.2 μm) was partially replaced with 3–8% MK (size 3.3 μm). The specimens were cured at 20 °C with 50% RH. The incorporation of 3% MK decreased the 7, 28 and 91 days compressive strength by ~24.3%, ~26.2% and 39.7%, respectively, whilst the incorporation of 5% MK decreased it by ~37.1%, ~10.8% and ~1.6%, respectively. The incorporation of 8% MK decreased the 7 and 28 days compressive strength by ~24.3% and ~4.6%, respectively, whilst the 91 days compressive strength was enhanced by 3.17%. Li et al. [30] found ~10% and ~36.7% reduction in the 1 day compressive strength of 50% slag/50% FA pastes activated with NaOH and sodium silicate solution by partially replacing slag with 5% and 10% MK, respectively. The incorporation of 5% and 10% MK decreased the 7 days compressive strength by ~3.2% and 9.3%, respectively, whilst the 28 days compressive strength was reduced by ~0% and ~9%, respectively. All the specimens were cured at 20 °C.

Table 4.2 Effect of MK on the compressive strength of FA geopolymers

References	Precursor	Precursor size/fineness	Activator	MK (%)	MK size/fineness	Type	Age (day)	Curing	Effect	Ratio
Alanazi et al. [20]	FA	–	NaOH and sodium silicate	5	–	Mortar	1, 3, 7 and 28	60 °C for 24 h	Increased	~30, ~42.8, ~45 and ~0
	FA	–	NaOH and sodium silicate	10	–	Mortar	1, 3, 7 and 28	60 °C for 24 h	Increased	~60, ~100, ~32.5 and ~6.75
Abbass et al. [21]	FA	37.92	NaOH and sodium silicate	5 and 20	16.05	Paste	56	20 °C with 70% RH	Increased	~163.6 and ~472.7
	FA	37.92	NaOH and sodium silicate	5 and 20	2.92	Paste	56	20 °C with 70% RH	Increased	~500 and ~518
Bheel et al. [22]	FA	379 m^2/kg	NaOH and sodium silicate solution	5, 10 and 15	18,000 m^2/kg	Concrete	28	75 °C for 2 h	Increased	~5.35, ~8.9 and ~2.3
	FA	379 m^2/kg	NaOH and sodium silicate solution	20	18,000 m^2/kg	Concrete	28	75 °C for 2 h	Decreased	~6.25
Duan et al. [23]	FA	51.09 μm	NaOH and sodium silicate	5, 10, 15 and 20	12.66 μm	Paste	28	20 °C with 90% RH	Increased	~9.6%, ~16.5%, ~28.7% and ~39.13

(continued)

Table 4.2 (continued)

References	Precursor	Precursor size/fineness	Activator	MK (%)	MK size/fineness	Type	Age (day)	Curing	Effect	Ratio
Guo et al. [24]	FA	D90: 3.34 μm	Phosphoric acid	20, 30 and 50	D90: 58.35 μm	Paste	7	40 °C with 90% RH for six days then at 80 °C for 24 h	Increased	292, 184.9 and 246.5
	FA	D90: 3.34 μm	Phosphoric acid	20, 30 and 50	D90: 58.35 μm	Paste	100	40 °C with 90% RH for six days then at 80 °C for 24 h	Increased	285.7, 307.9 and 590
Zhang et al. [25]	FA	37 μm	NaOH and waterglass	10, 30, 50, 70 and 90	8 μm	Paste	28	40 °C for 48 h	Increased	~325.53, ~437.2, ~368.1 and ~261.7 compared to those containing 10% MK
Gómez-Casero et al. [26]	FA	d50 19.87 μm	NaOH and sodium silicate	25, 50 and 75	d50 9.579 μm	Paste	7	60 °C with standard humidity for one day	Decreased	~22.2, ~8.3 and ~2.7
	FA	d50 19.87 μm	NaOH and sodium silicate	100	d50 9.579 μm	Paste	7	60 °C with standard humidity for one day	Increased	16.67

(continued)

Table 4.2 (continued)

References	Precursor	Precursor size/fineness	Activator	MK (%)	MK size/fineness	Type	Age (day)	Curing	Effect	Ratio
	FA	d50 19.87 μm	NaOH and sodium silicate	25, 50 and 75	d50 9.579 μm	Paste	28	60 °C with standard humidity for one day	Decreased	~31.2, ~20 and ~15.3
	FA	d50 19.87 μm	NaOH and sodium silicate	100	d50 9.579 μm	Paste	28	60 °C with standard humidity for one day	Increased	16.4
Fu et al. [27]	FA	19.3 μm	NaOH	20	1.8 μm	Paste	1, 3, 7 and 28	50 °C for 24 h	Decreased	~70.6, ~73.8, 75 and 77.3
	FA	19.3 μm	NaOH	50	1.8 μm	Paste	1, 3, 7 and 28	50 °C for 24 h	Decreased	~85.3, ~83.3, ~84 and ~85
	FA	19.3 μm	NaOH	100	1.8 μm	Paste	1, 3, 7 and 28	50 °C for 24 h	Decreased	~89.7, ~88.1, ~88.6 and 92.4
	FA	19.3 μm	KOH	20	1.8 μm	Paste	1, 3 and 7	50 °C for 24 h	Decreased	~11.11, ~8 and 0
	FA	19.3 μm	KOH	20	1.8 μm	Paste	28	50 °C for 24 h	Increased	~5.7
	FA	19.3 μm	KOH	50	1.8 μm	Paste	1, 3, 7 and 28	50 °C for 24 h	Decreased	44.4, ~52, ~48.4 and ~62.3
	FA	19.3 μm	KOH	100	1.8 μm	Paste	1, 3, 7 and 28	50 °C for 24 h	Decreased	~38.9, ~44, ~45 and 66
Zhang et al. [28]	FA	32 μm	KOH and potassium silicate	25, 50, 80 and 100	17 μm	Paste	7	20 °C with 90% RH	Increased	~97.4, ~250, ~229 and ~229

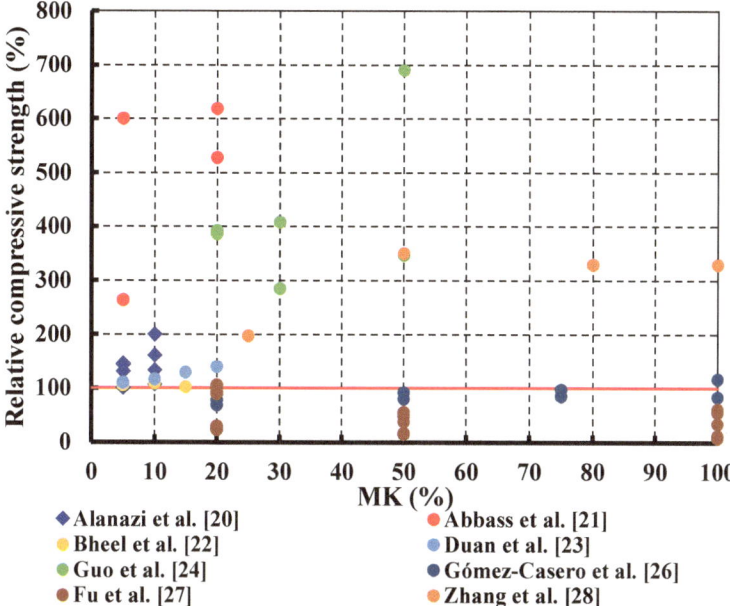

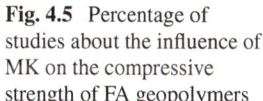

Fig. 4.4 Effect of MK on the compressive strength of FA geopolymers [20–24, 26–28]

Fig. 4.5 Percentage of studies about the influence of MK on the compressive strength of FA geopolymers

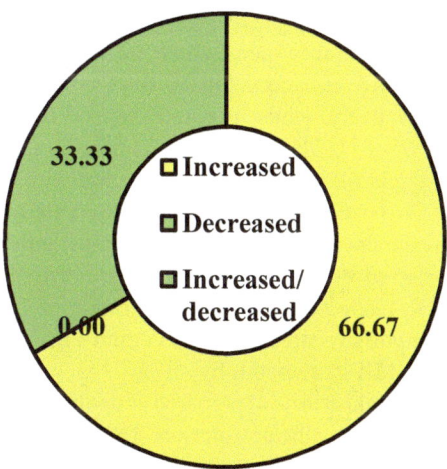

Sharmin et al. [31] prepared geopolymer mortars from 40% rice husk ash (RHA)/ 60% slag activated with NaOH and sodium silicate solution. The slag (fineness 405 m^2/kg) was partially replaced with 5–20% MK (fineness 4315.8 m^2/kg). The specimens were cured at 65 °C for 24 h. The 28 days compressive strength was enhanced by 52.45%, 57.66%, 62.5% and 91.94% with the incorporation of 5%, 10%, 15% and 20% MK, respectively. Rajamma et al. [32] prepared biomass FA

Fig. 4.6 Percentage of compressive studies of various media based on FA geopolymers containing MK

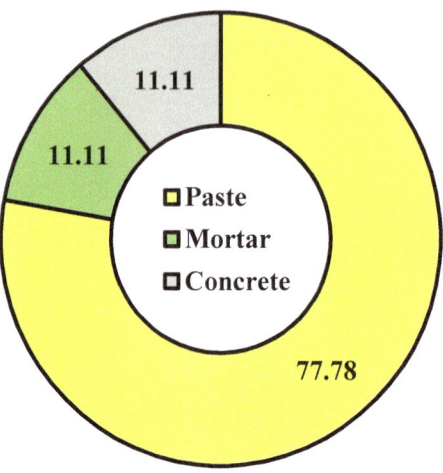

geopolymer pastes and mortars activated with NaOH and sodium silicate solution. Different concentrations of NaOH were used. The biomass FA (size 75 μm) was partially replaced with 20% and 40% MK. The specimens were cured at 60 °C with 95% RH for two days, then at 20 °C. The results showed 22.9% and 36.6% enhancement in the 10 days compressive strength of the pastes with the inclusion of 20% and 40% MK, respectively, when 10 M NaOH was used, whilst the strength was enhanced by 21.63% and 12.62%, respectively, when 18 M NaOH was used. For mortar specimens, when 10 M NaOH was used, the incorporation of 20% and 40% MK enhanced the 10 days compressive strength by 60.53% and 111.94%, respectively, whilst it was enhanced by 57.51% and 158.7%, respectively, when 18 M of NaOH was used. Mahmoodi et al. [33] prepared geopolymer pastes from ceramic tile waste (CTW) activated with NaOH and sodium silicate solution. The CTW (size 29.5 μm) was partially replaced with 15–45% MK (size 10.9 μm). The specimens were cured at room temperature. The results showed ~91%, ~45.6% and ~4% reduction in the 7 days compressive strength with including 15%, 30% and 45% MK, respectively. The incorporation of 15% and 30% MK decreased the 28 days compressive strength by ~65% and ~18.7%, respectively, whilst the incorporation of 45% MK increased it by ~3%.

Furlani et al. [34] prepared geopolymer pastes from steel slag activated with NaOH and sodium silicate solution. The steel slag (d50 = 36 or 70 μm) was replaced by 20–100% MK (size < 40 μm). The specimens were cured at room temperature for 24 h, then at 60 °C for an additional 24 h followed by room temperature. When fine slag (size 36 μm) was used, the 28 days compressive strength was enhanced by ~97%, ~110%, ~140%, ~130% and 100% with the incorporation of 20%, 40%, 60%, 80% and 100% MK, respectively, whilst it was enhanced by ~66%, ~90%, ~127%, ~121% and 100% when coarse slag (size 70 μm) was used, respectively. Bignozzi et al. [35] prepared geopolymer pastes from ladle slag (95% passing at 90 μm) activated with NaOH and sodium silicate solution. The ladle slag was partially replaced by 20–100% MK (95% passing at 30 μm). The specimens were cured at room temperature for 24 h at > 90% RH, then at 20 ± 2 °C with 55% RH for 6 days. The results showed ~420%, ~370%, ~390% and ~130% enhancement in the 7 days compressive strength with the inclusion of 20%, 30%, 40% and 50% MK, respectively, whilst the inclusion of 75% and 100% MK decreased it by ~1% and 70%, respectively. Rovnaník et al. [36] prepared waste brick powder (WBP) geopolymer pastes activated with NaOH and sodium silicate solution. The WBP (d50 = 8.5 μm) was replaced with 25–100% MK (d50 = 6.3 μm). The specimens were cured at room temperature for 4 h, then at 40 °C for 20 h, then in plastic bags under laboratory conditions. The results showed ~69.2%, 126.15%, ~246.1% and 176.9% enhancement in the 7 days compressive strength with the inclusion of 25%, 50%, 75% and 100% MK, respectively, whilst the 14 days compressive strength was enhanced by ~40%, ~38%, ~165% and ~90%, respectively. The results included in this part are summarized in Table 4.3.

Considering what has been reported thus far, it is clear that the findings on the effect of MK on the compressive strength of geopolymers based on unconventional or blended precursors are inconclusive (Fig. 4.7). This typically depends on precursor type/fineness, age of testing, activator type/concentration, Ca/Al ratio, Ca/Si ratio, Si/Al ratio, curing condition and MK ratio/fineness, but the type of precursor has a major effect. For example, when waste glass was used as the main precursor, the incorporation of MK generally decreased the compressive strength. This reduction could be attributed to the formation of Si–O–Al bonds that are weaker than Si–O–Si bonds found in the control. The incorporation of 3% MK showed the lowest compressive strength due to the high free alkali content [29]. The same trend was observed when CWP was used as the main precursor. Contrarily, when steel slag was used as the main precursor, the incorporation of MK increased the compressive strength at replacement ratios up to 100%. This could be attributed to the increase in the ratio of Al/Na with the inclusion of MK and the increase in porosity with increasing steel slag ratio [34]. Typically, the same trend was observed when biomass FA or WBP was used as the main precursor, of which the incorporation of MK up to 100% increased the compressive strength. Obviously, when RHA/slag was used as the main precursor, the incorporation of MK up to 20% increased the compressive strength. This could be attributed to the fast reaction between silica and alumina compared to the reaction of silica only [37]. When ladle slag was used as the main precursor, the incorporation of MK up to 50% increased the compressive strength, whilst 75% and 100% MK decreased it. In actually, the mentioned references in this section that focused on paste and mortar are roughly 66.67% and 33.33%, respectively (Fig. 4.8).

4.2 Flexural Strength and Splitting Tensile Strength

Li et al. [30] found ~22.5% and ~20% reduction in the 1 day flexural strength of 50% slag/50% FA pastes activated with NaOH and sodium silicate solution by partially replacing slag with 5% and 10% MK, respectively, whilst the 7 days flexural strength was enhanced by ~2.27% and ~13.64%, respectively. The incorporation of 5% MK

Table 4.3 Effect of MK on the compressive strength of other geopolymer types

References	Precursor	Precursor size/fineness	Activator	MK (%)	MK size/fineness	Type	Age (day)	Curing	Effect	Ratio
Pascual et al. [29]	Waste glass	10.2 μm	NaOH	3	3.3 μm	Mortar	7, 28 and 91	20 °C with 50% RH	Decreased	~24.3, ~26.2 and 39.7
	Waste glass	10.2 μm	NaOH	5	3.3 μm	Mortar	7, 28 and 91	20 °C with 50% RH	Decreased	~37.1, ~10.8 and ~1.6
	Waste glass	10.2 μm	NaOH	8	3.3 μm	Mortar	7 and 28	20 °C with 50% RH	Decreased	~24.3 and ~4.6
	Waste glass	10.2 μm	NaOH	8	3.3 μm	Mortar	91	20 °C with 50% RH	Increased	3.17
Li et al. [30]	Slag/FA	18.3 μm/48.1	NaOH and sodium silicate	5 and 10	69.4 μm	Paste	1	20 °C	Decreased	~10 and ~36.7
	Slag/FA	18.3 μm/48.1	NaOH and sodium silicate	5 and 10	69.4 μm	Paste	7	20 °C	Decreased	~3.2% and 9.3
	Slag/FA	18.3 μm/48.1	NaOH and sodium silicate	5 and 10	69.4 μm	Paste	28	20 °C	Increased	~0% and ~9
Sharmin et al. [31]	RHA/slag	2981.2/405 m²/kg	NaOH and sodium silicate	5, 10, 15 and 20	4315.8 m²/kg	Mortar	28	65 °C for 24 h	Increased	52.45, 57.66, 62.5 and 91.94
Rajamma et al. [32]	Biomass FA	75 μ	NaOH and sodium silicate	20 and 40	–	Paste	10	60 °C with 95% RH for two days	Increased	22.9 and 36.6 (10 M NaOH)
	Biomass FA	75 μ	NaOH and sodium silicate	20 and 40	–	Paste	10	60 °C with 95% RH for two days	Increased	21.63 and 12.62 (18 M NaOH)

(continued)

Table 4.3 (continued)

References	Precursor	Precursor size/fineness	Activator	MK (%)	MK size/fineness	Type	Age (day)	Curing	Effect	Ratio
	Biomass FA	75 μ	NaOH and sodium silicate	20 and 40	–	Mortar	10	60 °C with 95% RH for two days	Increased	60.53 and 111.94 (10 M NaOH)
	Biomass FA	75 μ	NaOH and sodium silicate	20 and 40	–	Mortar	10	60 °C with 95% RH for two days	Increased	57.51 and 158.7 (18 M NaOH)
Mahmoodi et al. [33]	CTW	29.5 μm	NaOH and sodium silicate	15, 30 and 45	10.9 μm	Paste	7	Room	Decreased	~91, ~45.6 and ~4
	CTW	29.5 μm	NaOH and sodium silicate	15 and 30	10.9 μm	Paste	28	Room	Decreased	~65 and ~18.7
	CTW	29.5 μm	NaOH and sodium silicate	45	10.9 μm	Paste	28	Room	Increased	~3
Furlani et al. [34]	Steel slag	d50 36 μm	NaOH and sodium silicate	20, 40, 60, 80 and 100	< 40 μm	Paste	28	Room for 24 h, then at 60 °C for 24 h	Increased	~97, ~110, ~140, ~130 and 100
		d50 70 μm	NaOH and sodium silicate	20, 40, 60, 80 and 100	< 40 μm	Paste	28	Room for 24 h, then at 60 °C for 24 h	Increased	~66, ~90, ~127, ~121 and 100
Bignozzi et al. [35]	Ladle slag	95% passing at 90 μm	NaOH and sodium silicate solution	20, 30, 40 and 50	95% passing at 30 μm	Paste	7	Room with > 90% RH, then at 20 ± 2 °C with 55% RH for 6 days	Increased	~420, ~370, ~390 and ~130
	Ladle slag	95% passing at 90 μm	NaOH and sodium silicate solution	75 and 100	95% passing at 30 μm	Paste	7	Room with > 90% RH, then at 20 ± 2 °C with 55% RH for 6 days	Decreased	~1 and 70

(continued)

Table 4.3 (continued)

References	Precursor	Precursor size/ fineness	Activator	MK (%)	MK size/ fineness	Type	Age (day)	Curing	Effect	Ratio
Rovnaník et al. [36]	WBP	d50 8.5 μm	NaOH and sodium silicate	25, 50, 75 and 100	d50 6.3 μm	Paste	7	4 h at room, then at 40 °C for 20 h	Increased	~69.2, 126.15, ~246.1 and 176.9
	WBP	d50 8.5 μm	NaOH and sodium silicate	25, 50, 75 and 100	d50 6.3 μm	Paste	14	4 h at room, then at 40 °C for 20 h	Increased	~40, ~38, ~165 and ~90

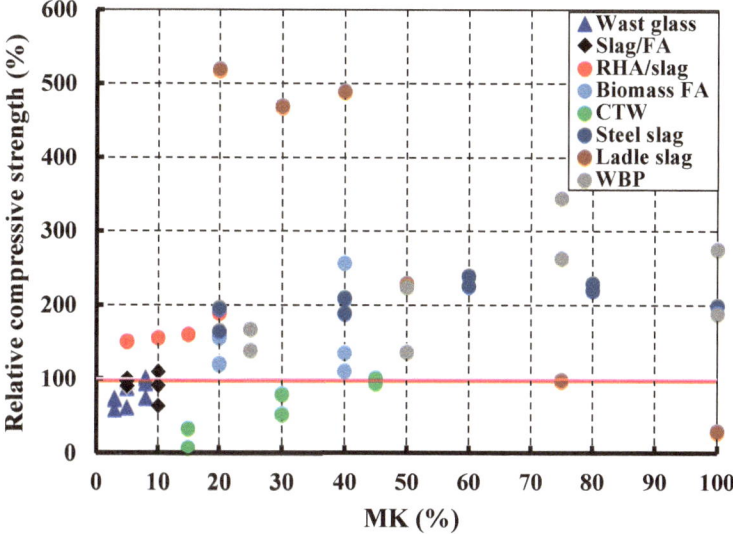

Fig. 4.7 Effect of MK on the compressive strength of geopolymers based on waste glass, slag/FA, RHA/slag, biomass FA, CTW and steel slag precursors [29–36]

Fig. 4.8 Percentage of compressive strength studies of various media of geopolymers based on waste glass, slag/FA, RHA/slag, biomass FA, CTW and steel slag precursors containing MK

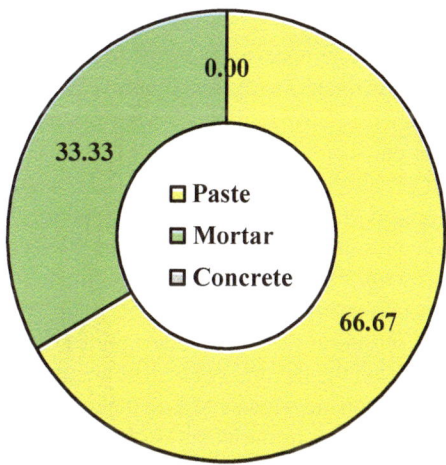

increased the 28 days flexural strength by ~4%, whilst 10% MK decreased it by ~2%. Bheel et al. [22] found ~8%, ~12.3% and ~3.7% enhancement in the 28 days flexural strength of self-compacting concrete specimens activated with NaOH and sodium silicate solution cured at 75 °C for 2 h by partially replacing FA (fineness 379 m^2/kg) with 5%, 10% and 15% MK (fineness 18,000 m^2/kg). They related this enhancement to the fine MK particles that can remove the microcracks and produce a more compact matrix. On the other hand, the incorporation of 20% MK decreased the strength by ~2.8%. Li et al. [3] found ~8%, ~24.3% and ~50% higher 1, 7 and 28 days flexural strength of AAS pastes activated with NaOH and sodium silicate solution by partially replacing slag (size 0.1–50 μm) with 10% MK (size 0.15–142 μm). The incorporation of 20% MK decreased the 1 day flexural strength by ~25%, whilst the 7 and 28 days flexural strength was enhanced by ~8% and ~47.5%, respectively. Bernal et al. [14] found 14.41% and 44.67% reduction in the 28 days flexural strength and splitting tensile strength of AAS concretes activated with NaOH and sodium silicate solution by partially replacing slag (fineness 498 m^2/kg) with 60% MK (fineness 391 m^2/kg), respectively. Buchwald et al. [18] prepared AAS pastes activated with NaOH. The slag (size 15.8 μm) was replaced with 25–100% MK (size 3.9 μm). The specimens were cured at 40 °C. The 28 days bending tensile strength was reduced by ~10%, ~35% and ~80% with the inclusion of 25%, 50% and 100% MK, respectively. They also found ~3.45%, ~9.2% and 1.1% enhancement in the 28 days splitting tensile strength of self-compacting concrete specimens activated with NaOH and sodium silicate solution cured at 75 °C for 2 h by partially replacing FA (fineness 379 m^2/kg) with 5%, 10% and 15% MK (fineness 18,000 m^2/kg). On the other hand, the incorporation of 20% MK decreased the strength by ~3.9%. Zhang et al. [28] prepared FA geopolymer pastes activated with KOH and potassium silicate solution. The FA (size 32 μm) was replaced with 25–100% MK (size 17 μm). The specimens were cured at 20 °C with 90% RH. The results showed ~200%, ~536.4%, ~590% and ~590% enhancement in the bending strength with the inclusion of 25%, 50%, 80% and 100% MK, respectively. Rovnaník et al. [36] found ~75%, ~250%, ~462% and 400% enhancement in the 7 days flexural strength of ladle slag geopolymer pastes activated with NaOH and sodium silicate solution with the incorporation of 25%, 50%, 75% and 100% MK, respectively, whilst the 14 days flexural strength was enhanced by ~22.2%, ~261%, ~372% and ~355%, respectively. Bheel et al. [22] found ~3.45%, ~9.2% and 1.1% enhancement in the 28 days splitting tensile strength of self-compacting concrete specimens activated with NaOH and sodium silicate solution cured at 75 °C for 2 h by partially replacing FA (fineness 379 m^2/kg) with 5%, 10% and 15% MK (fineness 18,000 m^2/kg). The results included in this part are summarized in Table 4.4.

Considering what has been reported thus far, it is clear that the findings on the effect of MK on the flexural strength of geopolymers are inconclusive (Fig. 4.9), whilst the findings on the effect of MK on the splitting tensile strength are too limited (Fig. 4.10). The variation in the flexural strength results typically depends on precursor type/fineness, age of testing, activator type/concentration, Ca/Al ratio, Ca/Si ratio, Si/Al ratio, curing condition and MK ratio/fineness, but the type of precursor has a major effect. For example, when slag was used as the main precursor, it was better to not increase the ratio of MK by more than 20% to obtain enhanced flexural strength. This enhancement could be attributed to the mitigation of microcracks with the inclusion of MK. The unreacted MK particles can act as a local autogenous shrinkage restraint [3] and increase binding capacity [38]. When FA was used as the main precursor, it was better to not increase the ratio of MK by more than 15% to obtain enhanced flexural strength. This enhancement could be attributed to the fine MK particles that can refine the pore size, increase packing and produce a denser microstructure [18] or

Table 4.4 Effect of MK on the flexural and splitting strengths of geopolymers

References	Precursor	Precursor size/fineness	Activator	MK (%)	MK size/fineness	Type	Age (day)	Curing	Effect	Ratio
Li et al. [30]	Slag/FA	18.3 μm/48.1	NaOH and sodium silicate	5 and 10	69.4 μm	Paste	1	20 °C	Decreased flexural	~22.5 and ~20
	Slag/FA	18.3 μm/48.1	NaOH and sodium silicate	5 and 10	69.4 μm	Paste	7	20 °C	Increased flexural	~2.27 and ~13.64
Bheel et al. [22]	FA	379 m^2/kg	NaOH and sodium silicate solution	5, 10 and 15	18,000 m^2/kg	Concrete	28	75 °C for 2 h	Increased flexural	~8%, ~12.3% and ~3.7
	FA	379 m^2/kg	NaOH and sodium silicate solution	20	18,000 m^2/kg	Concrete	28	75 °C for 2 h	Decreased flexural	~2.8
Bheel et al. [22]	FA	379 m^2/kg	NaOH and sodium silicate solution	5, 10 and 15	18,000 m^2/kg	Concrete	28	75 °C for 2 h	Increased splitting	~3.45, ~9.2 and 1.1
	FA	379 m^2/kg	NaOH and sodium silicate solution	20	18,000 m^2/kg	Concrete	28	75 °C for 2 h	Decreased splitting	~3.9
Li et al. [3]	Slag	0.1–50 μm	NaOH and sodium silicate	10	0.15–142 μm	Paste	1, 7 and 28	20 °C	Increased flexural	~8, ~24.3 and ~50

(continued)

Table 4.4 (continued)

References	Precursor	Precursor size/fineness	Activator	MK (%)	MK size/fineness	Type	Age (day)	Curing	Effect	Ratio
	Slag	0.1–50 μm	NaOH and sodium silicate	20	0.15–142 μm	Paste	1	20 °C	Decreased flexural	~25
	Slag	0.1–50 μm	NaOH and sodium silicate	20	0.15–142 μm	Paste	7 and 28	20 °C	Increased flexural	~8 and ~47.5
Bernal et al. [14]	Slag	498 m²/kg	NaOH and sodium silicate solution	60	391 m²/kg	Concrete	28	25 °C with >98% RH for 24 h, then at 27 °C with >90% RH	Decreased flexural	14.41
	Slag	498 m²/kg	NaOH and sodium silicate solution	60	391 m²/kg	Concrete	28	25 °C with >98% RH for 24 h, then at 27 °C with >90% RH	Decreased splitting	44.67
Buchwald et al. [18]	Slag	15.8 μm	NaOH	25, 50 and 100	3.9 μm	Paste	28	40 °C	Decreased bending	~10, ~35 and ~80
Zhang et al. [28]	FA	32 μm	KOH and potassium silicate	25, 50, 80 and 100	17 μm	Paste	7	20 °C with 90% RH	Increased bending	~200, ~536.4, ~590 and ~590
Rovnanik et al. [36]	WBP	d50 8.5 μm	NaOH and sodium silicate	25, 50, 75 and 100	d50 6.3 μm	Paste	7	4 h at room, then at 40 °C for 20 h	Increased flexural	~75, ~250, ~462 and 400

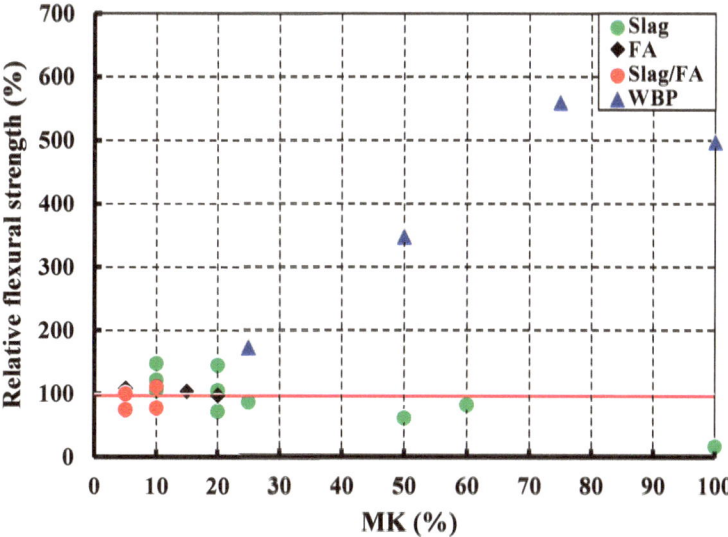

Fig. 4.9 Effect of MK on the flexural strength of geopolymers based on slag, FA, slag/FA and WBP precursors [3, 14, 18, 22, 28, 30, 36]

to the fine MK particles that can remove the microcracks and produce a more compact microstructure [22]. When WBP was used as the main precursor, the flexural strength increased as the MK ratio increased. This enhancement could be attributed to the synergistic impact of MK on the slow reaction of WBP [36]. Regarding splitting tensile strength, although there are only two available studies, it is better to not increase the ratio of MK by more than 15% to obtain enhanced splitting strength. This enhancement could be attributed to the fine MK particles that can refine the pore size, increase packing and produce a denser microstructure [22]. In actuality, the mentioned references in this section (related to flexural strength) that focused on paste and concrete are roughly 71.43% and 28.57%, respectively (Fig. 4.11).

4.3 Elastic Modulus

Li et al. [30] found ~23.6% enhancement in the 1 day elastic modulus of AAS mortars activated with NaOH and sodium silicate solution by partially replacing slag with 5% MK, whilst partially replacing slag with 10% MK did not show a fundamental change. At the age of 7 days, the incorporation of 5% and 10% MK increased the elastic modulus by ~7.8% and ~8.47%, respectively. Alanazi et al. [20] found ~42.8%, ~23.1% and ~11.1% enhancement in the 1, 3 and 7 days elastic modulus of FA mortars activated with NaOH and sodium silicate solution by partially replacing FA with 5% MK, respectively, when sodium silicate/NaOH was 1, whilst

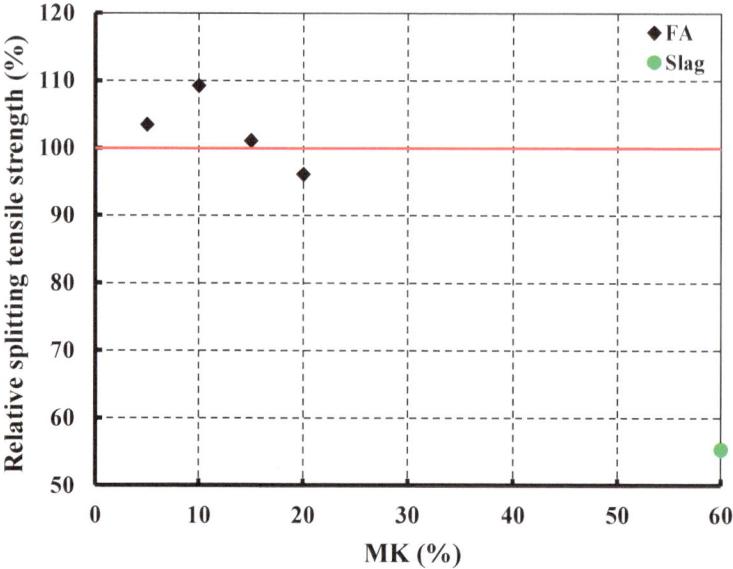

Fig. 4.10 Effect of MK on the splitting tensile strength of geopolymers based on slag and FA precursors [14, 22]

Fig. 4.11 Percentage of flexural strength studies of various media of geopolymers based on slag, FA, slag/FA and WBP precursors containing MK

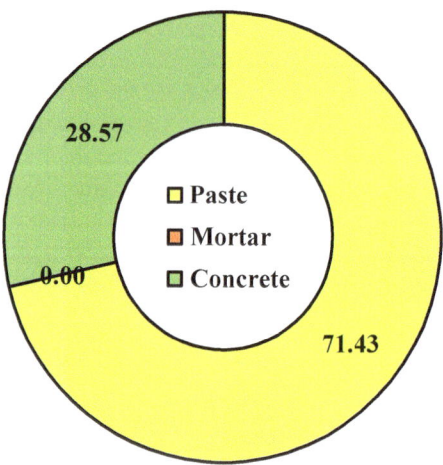

the 28 days elastic modulus was reduced by ~11.1%. The incorporation of 10% MK increased the 1, 3, 7 and 28 days elastic modulus by ~114.3%, ~69.23%, ~0% and ~8.3%, respectively. When sodium silicate/NaOH was 2.5, the incorporation of 5% MK increased the 1, 3, 7 and 28 days elastic modulus by ~50%, ~38.5%, 66.7% and ~3.2%, respectively, whilst the incorporation of 10% MK increased it by ~65%, ~7.7%, ~94.4% and ~19.4%, respectively. Considering what has been reported thus far, it is difficult to reach a conclusion about the effect of MK on the elastic modulus of geopolymers because there are only two available studies in this section.

References

1. G.F. Huseien, J. Mirza, M. Ismail, S. Ghoshal, M.A.M. Ariffin, Effect of metakaolin replaced granulated blast furnace slag on fresh and early strength properties of geopolymer mortar. Ain Shams Eng. J. **9**(4), 1557–1566 (2018)
2. A.M. Rashad, Additives to increase carbonation resistance of slag activated with sodium sulfate. ACI Mater. J. **119**(2) (2022)
3. Z. Li, M. Nedeljković, B. Chen, G. Ye, Mitigating the autogenous shrinkage of alkali-activated slag by metakaolin. Cem. Concr. Res. **122**, 30–41 (2019)
4. S.A. Bernal, J.L. Provis, V. Rose, R.M. De Gutierrez, Evolution of binder structure in sodium silicate-activated slag-metakaolin blends. Cement Concr. Compos. **33**(1), 46–54 (2011)
5. S.A. Bernal, R.M. De Gutierrez, J.L. Provis, V. Rose, Effect of silicate modulus and metakaolin incorporation on the carbonation of alkali silicate-activated slags. Cem. Concr. Res. **40**(6), 898–907 (2010)
6. S.A. Bernal, R.M. De Gutiérrez, J.L. Provis, Engineering and durability properties of concretes based on alkali-activated granulated blast furnace slag/metakaolin blends. Constr. Build. Mater. **33**, 99–108 (2012)
7. S.A. Bernal, Effect of the activator dose on the compressive strength and accelerated carbonation resistance of alkali silicate-activated slag/metakaolin blended materials. Constr. Build. Mater. **98**, 217–226 (2015)
8. B. El Moustapha, S. Bonnet, A. Khelidj, N. Leklou, D. Froelich, I.A. Babah, C. Charbuillet, A. Khalifa, Compensation of the negative effects of micro-encapsulated phase change materials by incorporating metakaolin in geopolymers based on blast furnace slag. Constr. Build. Mater. **314**, 125556 (2022)
9. Z. Chen, H. Ye, Improving sulphuric acid resistance of slag-based binders by magnesium-modified activator and metakaolin substitution. Cement Concr. Compos. **131**, 104605 (2022)
10. M.A. Asaad, G.F. Huseien, R.P. Memon, S. Ghoshal, H. Mohammadhosseini, R. Alyousef, Enduring performance of alkali-activated mortars with metakaolin as granulated blast furnace slag replacement. Case Stud. Constr. Mater. **16**, e00845 (2022)
11. M. Gharieb, Y.A. Mosleh, A.M. Rashad, Properties and corrosion behaviour of applicable binary and ternary geopolymer blends. Int. J. Sustain. Eng. **14**(5), 1068–1080 (2021)
12. V.V. Praveen Kumar, N. Prasad, S. Dey, Influence of metakaolin on strength and durability characteristics of ground granulated blast furnace slag based geopolymer concrete. Struct. Concr. **21**(3), 1040–1050 (2020)
13. M.G. Khalil, F. Elgabbas, M.S. El-Feky, H. El-Shafie, Performance of geopolymer mortar cured under ambient temperature. Constr. Build. Mater. **242**, 118090 (2020)
14. S. Bernal, R.M. de Gutiérrez, F. Ruiz, H. Quiñones, J. Provis, High-temperature performance of mortars and concretes based on alkali-activated slag/metakaolin blends. Mater. Constr. **62**(308), 471–488 (2012)
15. O. Burciaga-Díaz, J.I. Escalante-García, R. Arellano-Aguilar, A. Gorokhovsky, Statistical analysis of strength development as a function of various parameters on activated metakaolin/slag cements. J. Am. Ceram. Soc. **93**(2), 541–547 (2010)
16. O. Burciaga-Díaz, L.Y. Gómez-Zamorano, J.I. Escalante-García, Influence of the long term curing temperature on the hydration of alkaline binders of blast furnace slag-metakaolin. Constr. Build. Mater. **113**, 917–926 (2016)
17. H. Peng, C. Cui, Z. Liu, C. Cai, Y. Liu, Synthesis and reaction mechanism of an alkali-activated metakaolin-slag composite system at room temperature. J. Mater. Civ. Eng. **31**(1), 04018345 (2019)

18. A. Buchwald, H. Hilbig, C. Kaps, Alkali-activated metakaolin-slag blends—performance and structure in dependence of their composition. J. Mater. Sci. **42**(9), 3024–3032 (2007)
19. O. Burciaga-Díaz, J.I. Escalante-García, Comparative performance of alkali activated slag/metakaolin cement pastes exposed to high temperatures. Cement Concr. Compos. **84**, 157–166 (2017)
20. H. Alanazi, J. Hu, Y.-R. Kim, Effect of slag, silica fume, and metakaolin on properties and performance of alkali-activated fly ash cured at ambient temperature. Constr. Build. Mater. **197**, 747–756 (2019)
21. A.M. Abbass, R. Firdous, J.N. Yankwa Djobo, D. Stephan, M.A. Elrahman, The role of chemistry and fineness of metakaolin on the fresh properties and heat resistance of blended fly ash-based geopolymer. SN Appl. Sci. **5**(5), 1–17 (2023)
22. N. Bheel, P. Awoyera, T. Tafsirojjaman, N.H. Sor, Synergic effect of metakaolin and groundnut shell ash on the behavior of fly ash-based self-compacting geopolymer concrete. Constr. Build. Mater. **311**, 125327 (2021)
23. P. Duan, C. Yan, W. Zhou, Influence of partial replacement of fly ash by metakaolin on mechanical properties and microstructure of fly ash geopolymer paste exposed to sulfate attack. Ceram. Int. **42**(2), 3504–3517 (2016)
24. H. Guo, B. Zhang, L. Deng, P. Yuan, M. Li, Q. Wang, Preparation of high-performance silico-aluminophosphate geopolymers using fly ash and metakaolin as raw materials. Appl. Clay Sci. **204**, 106019 (2021)
25. X. Zhang, X. Zhang, X. Li, M. Ma, Z. Zhang, X. Ji, Slurry rheological behaviors and effects on the pore evolution of fly ash/metakaolin-based geopolymer foams in chemical foaming system with high foam content. Constr. Build. Mater. **379**, 131259 (2023)
26. M. Gómez-Casero, C. De Dios-Arana, J. Bueno-Rodríguez, L. Pérez-Villarejo, D. Eliche-Quesada, Physical, mechanical and thermal properties of metakaolin-fly ash geopolymers. Sustain Chem Pharm **26**, 100620 (2022)
27. C. Fu, H. Ye, K. Zhu, D. Fang, J. Zhou, Alkali cation effects on chloride binding of alkali-activated fly ash and metakaolin geopolymers. Cement Concr. Compos. **114**, 103721 (2020)
28. H.Y. Zhang, V. Kodur, B. Wu, L. Cao, S.L. Qi, Comparative thermal and mechanical performance of geopolymers derived from metakaolin and fly ash. J. Mater. Civ. Eng. **28**(2), 04015092 (2016)
29. A.B. Pascual, T.M. Tognonvi, A. Tagnit-Hamou, Optimization study of waste glass powder-based alkali activated materials incorporating metakaolin: activation and curing conditions. J. Clean. Prod. **308**, 127435 (2021)
30. Z. Li, X. Liang, Y. Chen, G. Ye, Effect of metakaolin on the autogenous shrinkage of alkali-activated slag-fly ash paste. Constr. Build. Mater. **278**, 122397 (2021)
31. A. Sharmin, U.J. Alengaram, M.Z. Jumaat, M.O. Yusuf, S.A. Kabir, I.I. Bashar, Influence of source materials and the role of oxide composition on the performance of ternary blended sustainable geopolymer mortar. Constr. Build. Mater. **144**, 608–623 (2017)
32. R. Rajamma, J.A. Labrincha, V.M. Ferreira, Alkali activation of biomass fly ash–metakaolin blends. Fuel **98**, 265–271 (2012)
33. O. Mahmoodi, H. Siad, M. Lachemi, S. Dadsetan, M. Sahmaran, Development of ceramic tile waste geopolymer binders based on pre-targeted chemical ratios and ambient curing. Constr. Build. Mater. **258**, 120297 (2020)
34. E. Furlani, S. Maschio, M. Magnan, E. Aneggi, F. Andreatta, M. Lekka, A. Lanzutti, L. Fedrizzi, Synthesis and characterization of geopolymers containing blends of unprocessed steel slag and metakaolin: the role of slag particle size. Ceram. Int. **44**(5), 5226–5232 (2018)
35. M.C. Bignozzi, S. Manzi, I. Lancellotti, E. Kamseu, L. Barbieri, C. Leonelli, Mix-design and characterization of alkali activated materials based on metakaolin and ladle slag. Appl. Clay Sci. **73**, 78–85 (2013)
36. P. Rovnaník, P. Rovnanikova, M. Vyšvařil, S. Grzeszczyk, E. Janowska-Renkas, Rheological properties and microstructure of binary waste red brick powder/metakaolin geopolymer. Constr. Build. Mater. **188**, 924–933 (2018)

37. P. De Silva, K. Sagoe-Crenstil, V. Sirivivatnanon, Kinetics of geopolymerization: role of Al_2O_3 and SiO_2. Cem. Concr. Res. **37**(4), 512–518 (2007)
38. Z. Chen, H. Ye, Influence of metakaolin and limestone on chloride binding of slag activated by mixed magnesium oxide and sodium hydroxide. Cement Concr. Compos.Concr. Compos. **127**, 104397 (2022)

Chapter 5
Effect of Metakaolin on the Durability of Geopolymers

5.1 Carbonation and Corrosion Resistance

Rashad [1] reported that the incorporation of 10% MK into slag pastes activated with Na_2SO_4 increased the carbonation depth by 66.67% and 10.6% after exposure to accelerated carbonation (5% CO_2) for 2 and 4 weeks, respectively, whilst the carbonation depth was reduced by 30.38% after 8 weeks of exposure. Bernal et al. [2] found a higher carbonation depth of AAS mortars activated with NaOH and sodium silicate solution by partially replacing slag with 10% and 20% MK. Bernal [3] reported that the incorporation of 10% and 20% MK into alkali-activated slag (AAS) mortars activated with NaOH and sodium silicate solution reduced the carbonation depth after exposure to accelerated carbonation ($1 \pm 0.2\%$ CO_2) for 340 and 540 h when high ratios of SiO_2/Al_2O_3 were used. Asaad et al. [4] found 5.5% reduction in the carbonation depth of AAS mortars activated with NaOH and sodium silicate solution after exposure to CO_2 for 28 days with the inclusion of 10% MK, whilst the inclusion of 15% and 25% MK increased it by 4.11% and 10.96%, respectively. Gharieb et al. [5] found a lower corrosion rate (0.653 µm/year) of steel rebars embedded into AAS pastes activated with NaOH and sodium silicate solution by partially replacing slag (average diameter = 37.183 µm) with 30% MK (average diameter = 4.507 µm) compared to those free from MK (corrosion rate 1.475 µm/year). Considering what has been reported thus far, it is difficult to reach a conclusion due to the limited available studies and the conflicting results.

5.2 Other Durability Aspects

Alanazi et al. [6] found lower surface resistivity of fly ash (FA) geopolymer mortars activated with NaOH and sodium silicate solution cured at ambient temperature by partially replacing FA with 5% and 10% MK. Duan et al. [7] found higher sulphate resistance of FA pastes activated with NaOH and sodium silicate solution by partially replacing FA (d50 51.09 μm) with 5–20% MK (d50 12.66 μm) after exposure to 5% sodium sulphate for 28, 90 and 180 days. Bernal et al. [8] prepared AAS concrete specimens activated with NaOH and sodium silicate solution. The slag (size 0.1–74 μm d50 15 μm) was partially replaced with 10% and 20% MK (size 1.8–100 μm d50 12.2 μm). Different ratios of Si/Al were used. The specimens were cured at 25 ± 5 °C with 90% RH. The 28 days rapid chloride permeability was decreased with the inclusion of 10% and 20% MK. When the Si/Al ratio was 3.6, the reduction in the total charge passed was 40% and 30% with the incorporation of 10% and 20% MK, respectively. The 90 days chloride passed was slightly reduced with the inclusion of MK. El Moustapha et al. [9] found an increase in the chloride binding capacity of AAS mortars activated with NaOH and sodium silicate solution with the incorporation of 10% and 20% MK. Shi et al. [10] found ~44.4%, ~73.3%, 88.9% and ~97.24% reduction in the expansion caused by the alkali–silica reaction of AAS mortars activated by waterglass after exposure to 1 mol/L NaOH for 14 days by partially replacing slag with 10%, 30%, 50% and 70% MK, whilst the expansion was decreased by ~57.9%, ~73.7%, ~86.8% and 98.2% after 28 days of exposure, respectively.

Chen and Ye [11] found higher acid resistance of AAS pastes activated with NaOH and MgO with the incorporation of 20% MK after exposure to 1–5% H_2SO_4 for up to 63 days. They related this improvement to denser microstructure and the formation of silica-rich gel which can limit acid penetration. Asaad et al. [4] found higher acid resistance of AAS mortars activated with NaOH and sodium silicate solution after exposure to 10% H_2SO_4 for 180 and 360 days by partially replacing slag with 5- 25% MK. After 180 days of exposure, the compressive strength loss of the control specimens was 42.3%, whilst it was 29.7% for those containing 25% MK. After 360 days of exposure, the strength loss of the control was 68.1%, whilst it was 52.6% for those containing 25% MK. They related this improvement to the restriction of the formulation of gypsum which reduced the internal cracks. They also found an enhancement in the resistance to wet/dry cycles of AAS mortars with the incorporation of 5% and 10% MK. After 50 cycles, the compressive strength loss was reduced by 1.22% and 0.96% with the incorporation of 5% and 10% MK, respectively. After 100 cycles, the strength loss of the control specimens was reduced from 3.62% to 2.91% with the incorporation of 10% MK. After 150 cycles, the strength loss of the control specimens was reduced from 6.74% to 5.96% and 5.08% with the incorporation of 5% and 10% MK, respectively, whilst the incorporation of 15%, 20% and 25% MK increased it by 7.23%, 7.28% and 7.88%, respectively.

Abbass et al. [12] reported that FA geopolymer pastes activated with NaOH and sodium silicate solution containing 5% and 20% MK showed higher compressive strength after exposure to 90 °C for 2 h. The compressive strength started to decrease after exposure to 300 °C, whilst it reached its lowest value after exposure to 500 °C. Increasing the MK fineness and alumina ratio led to increasing fire resistance. Zhang et al. [13] reported that the incorporation of MK into FA geopolymer pastes activated with KOH and potassium silicate solution has a negative effect on the residual compressive strength after exposure to 500 °C. After exposure to 500 °C, the residual compressive strength of the specimens containing 50% MK was 44%, whilst it was

92% for the control. They related this reduction to the higher mass loss with this inclusion of MK. Burciaga-Díaz et al. [14] found higher residual compressive strength of AAS pastes activated with NaOH and sodium silicate solution after exposure to temperatures up to 1000 °C by partially replacing slag (fineness 465.3 m^2/kg) with 50% MK (fineness 946.9 m^2/kg). The results obtained by Bernal et al. [15] revealed that the residual compressive strength of AAS mortars activated with NaOH and sodium silicate solution after exposure to 600 °C is comparable to that containing 60% and 80% MK. After exposure, the permeable volumes of the pores of the specimens containing 0%, 20%, 40%, 60% and 80% MK were increased by 87.92%, 161.7%, 107.4%, 95.4% and 82.83%, respectively. They also found comparable relative residual strength of AAS concretes after exposure to 400, 800 and 1000 °C to those containing 60% MK. The results included in this section are summarized in Table 5.1.

Considering what has been reported thus far, it is clear that the incorporation of 5–10% MK into FA mortars decreased surface resistivity due to its high aluminium ratio. Higher aluminium led to high conductivity and low electrical resistivity [6]. The incorporation of 5–20% MK into FA pastes improved sulphate resistance. This improvement could be attributed to the high Al_2O_3 and SiO_2 contents in MK. The incorporation of MK can increase the solid raw materials reactivity as well as the solid particles contact surface area with the alkaline solution [7]. The incorporation of 10% and 20% MK into AAS concretes decreased the rapid chloride permeability due to the reduction in the capillarity with including MK [8]. The incorporation of MK up to 70% into AAS mortars decreased the expansion of the alkali–silica reaction due to the high aluminium, low calcium and low pore solution alkalinity [10]. With high efficiency, up 25% of MK can be used to increase the acid resistance of AAS matrices. This improvement could be related to the denser microstructure and the formation of silica-rich gel with the inclusion of MK which can limit the acid penetration [11]. Asaad et al. [4] related this improvement to the restriction of the formulation of gypsum with the inclusion of MK which reduced the internal cracks. Although MK geopolymers can resist elevated temperatures [16, 17], the incorporation of MK into FA or slag matrices exhibited an adverse effect on the residual strength after exposure to elevated temperatures as reported in [13, 14] or did not affect the residual strength of AAS mortars as reported in [14, 15]. Due to this inconsistency in the results, it is recommended to increase the number of studies on this point to reach a clear conclusion.

Table 5.1 Effect of MK on various durability aspects of geopolymers

References	Precursor	Precursor size/fineness	Activator	MK (%)	MK size/fineness	Type	Curing	Effect
Alanazi et al. [6]	FA	–	NaOH and sodium silicate	5 and 10	–	Mortar	Room	– Decreased surface resistivity
Duan et al. [7]	FA	51.09 μm	NaOH and sodium silicate	5–20	12.66 μm	Paste	20 °C with 90% RH	– Increased sulphate resistance
Bernal et al. [8]	Slag	0.1–74 μm	NaOH and sodium silicate	10 and 20	1.8–100 μm	Concrete	25 ± 5 °C with 90% RH	– Decreased rapid chloride permeability
El Moustapha et al. [9]	Slag	440 m²/kg	NaOH and sodium silicate	10 and 20	170 m²/kg	Mortar	20 °C with 50% RH	– Increased chloride binding capacity
Shi et al. [10]	Slag	–	Waterglass	10–70	–	Mortar	20 °C ± 2 with 90% RH for 24 h, then steam at 80 °C for 22 h	– Decreased alkali–silica reaction expansion
Chen and Ye [11]	Slag	501 m²/kg	NaOH and MgO	20	> 501 m²/kg	Paste	23 °C ± 5 with 90% RH	– Increased acid resistance
Asaad et al. [4]	Slag	12 μm	NaOH and sodium silicate	5–25	12 μm	Mortar	27 °C with 75% RH	– Increased acid resistance

(continued)

Table 5.1 (continued)

References	Precursor	Precursor size/fineness	Activator	MK (%)	MK size/fineness	Type	Curing	Effect
Abbass et al. [12]	FA	37.92	NaOH and sodium silicate	5 and 20	16.05 2.92	Paste	20 °C with 70% RH	– Increased compressive strength after exposure to 90 °C, then decreased after exposure to 300 and 500 °C – As fineness and alumina content the content increased in MK as the fire resistance increased
Zhang et al. [13]	FA	32 μm	KOH and potassium silicate	50	17 μm	Paste	20 °C with 90% RH	– Decreased residual strength after exposure to 500 °C
Burciaga-Díaz et al. [14]	Slag	465.3 m^2/g	NaOH and sodium silicate	50	946.9 m^2/g	Paste	20 °C with 80% RH for 24 h, then at 60 °C for 48 h	– Decreased residual strength after exposure to up to 1000 °C
Bernal et al. [15]	Slag	498 m^2/kg	NaOH and sodium silicate solution	40	391 m^2/kg	Mortar	25 °C with > 98% RH for 24 h, then at 27 °C with > 90% RH	– No effect after exposure to 600°C
	Slag	498 m^2/kg	NaOH and sodium silicate solution	60	391 m^2/kg	Mortar	25 °C with > 98% RH for 24 h, then at 27 °C with > 90% RH	– No effect after exposure to 400–1000 °C

References

1. A.M. Rashad, Additives to increase carbonation resistance of slag activated with sodium sulfate. ACI Mater. J. **119**(2) (2022)
2. S.A. Bernal, R.M. De Gutierrez, J.L. Provis, V. Rose, Effect of silicate modulus and metakaolin incorporation on the carbonation of alkali silicate-activated slags. Cem. Concr. Res. **40**(6), 898–907 (2010)
3. S.A. Bernal, Effect of the activator dose on the compressive strength and accelerated carbonation resistance of alkali silicate-activated slag/metakaolin blended materials. Constr. Build. Mater. **98**, 217–226 (2015)
4. M.A. Asaad, G.F. Huseien, R.P. Memon, S. Ghoshal, H. Mohammadhosseini, R. Alyousef, Enduring performance of alkali-activated mortars with metakaolin as granulated blast furnace slag replacement. Case Stud. Constr. Mater. **16**, e00845 (2022)
5. M. Gharieb, Y.A. Mosleh, A.M. Rashad, Properties and corrosion behaviour of applicable binary and ternary geopolymer blends. Int. J. Sustain. Eng. **14**(5), 1068–1080 (2021)
6. H. Alanazi, J. Hu, Y.-R. Kim, Effect of slag, silica fume, and metakaolin on properties and performance of alkali-activated fly ash cured at ambient temperature. Constr. Build. Mater. **197**, 747–756 (2019)
7. P. Duan, C. Yan, W. Zhou, Influence of partial replacement of fly ash by metakaolin on mechanical properties and microstructure of fly ash geopolymer paste exposed to sulfate attack. Ceram. Int. **42**(2), 3504–3517 (2016)
8. S.A. Bernal, R.M. De Gutiérrez, J.L. Provis, Engineering and durability properties of concretes based on alkali-activated granulated blast furnace slag/metakaolin blends. Constr. Build. Mater. **33**, 99–108 (2012)
9. B. El Moustapha, S. Bonnet, A. Khelidj, N. Maranzana, D. Froelich, A. Khalifa, I.A. Babah, Effects of microencapsulated phase change materials on chloride ion transport properties of geopolymers incorporating slag and, metakaolin, and cement-based mortars. J. Build. Eng. 106887 (2023)
10. Z. Shi, C. Shi, J. Zhang, S. Wan, Z. Zhang, Z. Ou, Alkali-silica reaction in waterglass-activated slag mortars incorporating fly ash and metakaolin. Cem. Concr. Res. **108**, 10–19 (2018)
11. Z. Chen, H. Ye, Improving sulphuric acid resistance of slag-based binders by magnesium-modified activator and metakaolin substitution. Cement Concr. Compos. **131**, 104605 (2022)
12. A.M. Abbass, R. Firdous, J.N. Yankwa Djobo, D. Stephan, M.A. Elrahman, The role of chemistry and fineness of metakaolin on the fresh properties and heat resistance of blended fly ash-based geopolymer. SN Appl. Sci. **5**(5), 1–17 (2023)
13. H.Y. Zhang, V. Kodur, B. Wu, L. Cao, S.L. Qi, Comparative thermal and mechanical performance of geopolymers derived from metakaolin and fly ash. J. Mater. Civ. Eng. **28**(2), 04015092 (2016)
14. O. Burciaga-Díaz, J.I. Escalante-García, Comparative performance of alkali activated slag/metakaolin cement pastes exposed to high temperatures. Cement Concr. Compos. **84**, 157–166 (2017)
15. S. Bernal, R.M. de Gutiérrez, F. Ruiz, H. Quiñones, J. Provis, High-temperature performance of mortars and concretes based on alkali-activated slag/metakaolin blends. Mater. Constr. **62**(308), 471–488 (2012)
16. A.M. Rashad, A.A. Hassan, S.R. Zeedan, An investigation on alkali-activated Egyptian metakaolin pastes blended with quartz powder subjected to elevated temperatures. Appl. Clay Sci. **132**, 366–376 (2016)
17. A.M. Rashad, A.S. Ouda, Thermal resistance of alkali-activated metakaolin pastes containing nano-silica particles. J. Therm. Anal. Calorim. **136**, 609–620 (2019)

Chapter 6
General Viewpoint and Ideas for Future Work

All things considered, the introduction of MK increased the workability when slag was used as the precursor due to increasing aluminium and silicate amounts and reducing calcium amount. Contrarily, the introduction of MK into FA mixtures decreased the workability due to increasing irregular spiny plate-like MK particles and reducing spherical FA particles. Regardless the type of precursor, all studies agreed that the introduction of MK decelerated the setting time due to the slow process of polycondensation. The introduction of MK increased the density of AAS specimens, due to the formation of N–A–S–H gel from MK in addition to C–S–H gel, as well as fly ash (FA) specimens, due to the higher density of FA/MK compared to that of FA alone. Contrarily, the introduction of MK decreased the density of ladle slag due to the reduction in the specific gravity. To be clearer, the compressive strength data of geopolymers based on slag precursor (Fig. 4.1), FA precursor (Fig. 4.4) and other precursors (Fig. 4.7) are plotted in one Fig. 6.1. One should be informed that, despite contradictory data about the influence of MK on the compressive strength of geopolymers, almost always, the introduction of MK up to 30%, 30%, 20% and 50% may improve the compressive strength of slag, FA, rice husk ash (RHA)/slag and ladle slag geopolymers, respectively. Regardless the type of precursor, age of testing, MK content, activator type/concentration and curing condition, considerable earlier studies (around 39.47%) reported higher compressive strength with the introduction of MK, whilst around 28.95% reported lower compressive strength. Withal, around 31.58% of them reported that the introduction of MK may increase or decrease the compressive strength (Fig. 6.2). This typically depends on precursor type/fineness, age of testing, activator type/concentration, Ca/Al ratio, Ca/Si ratio, Si/Al ratio, curing condition and MK ratio/fineness. Figure 6.3 shows that around 54.05% and 27.03% of the prior studies added MK to the slag precursor and FA precursor, respectively, to determine out its effect on the compressive strength, whilst less attention focused on other types of precursors such as slag/FA, RHA/ slag, waste glass, ceramic tile waste (CTW), steel slag and ladle slag. Regarding flexural strength, to reach acceptable results, it is recommended to not increase the

A. M. Rashad, *Metakaolin Effect on Geopolymers' Properties*,
SpringerBriefs in Applied Sciences and Technology,
https://doi.org/10.1007/978-3-031-45151-5_6

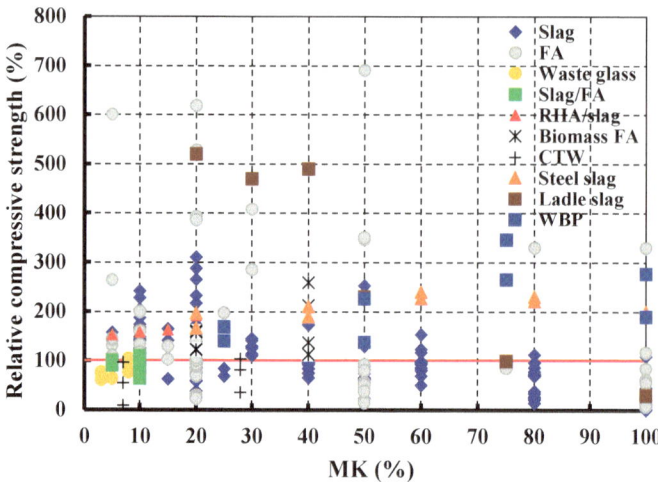

Fig. 6.1 Compressive strength variation of different geopolymer types with introducing MK [1–35]

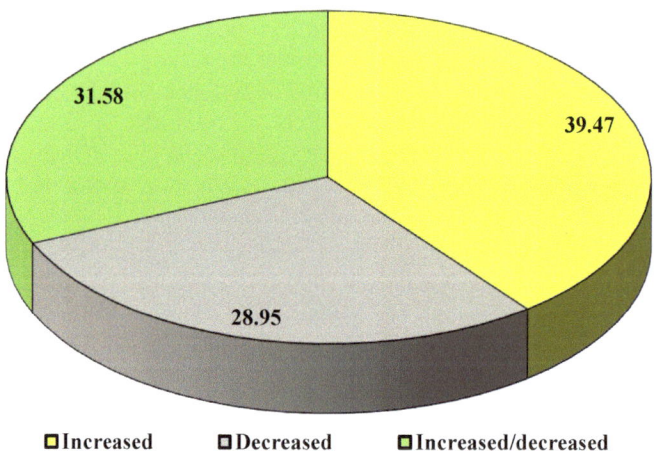

Fig. 6.2 Effect of MK on the compressive strength of geopolymers based on different precursors

MK ratio by more than 20% and 15% in slag and FA geopolymers, respectively, whilst the flexural strength increased with increasing the MK ratio in waste brick powder (WBP) geopolymers. Similarly, it is recommended to not increase the MK ratio by more than 15% in slag and FA geopolymers to obtain acceptable splitting tensile strength. It was found that the incorporation of 5% MK into AAS mortars or FA geopolymer mortars can enhance the elastic modulus.

The introduction of MK into slag geopolymers has a negative effect on the porosity as confirmed by around 63.6% of the earlier results. However, to obtain reduced porosity, it is recommended that the ratio of MK not exceed 10%. The introduction

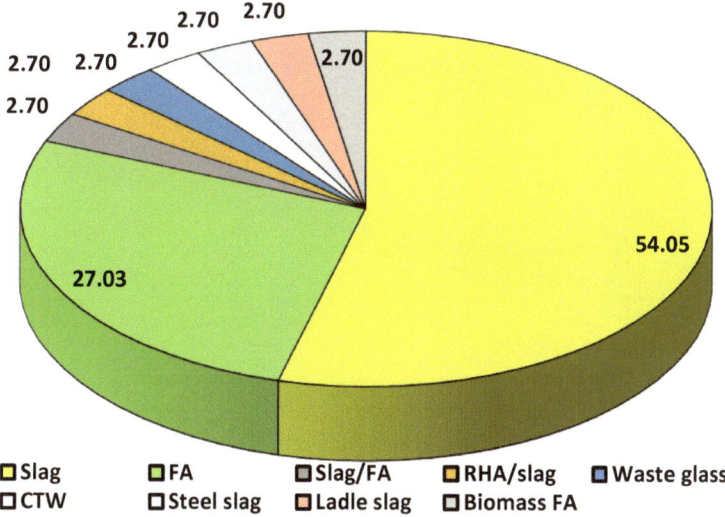

2.70 2.70
2.70
2.70 2.70
2.70
2.70

54.05

27.03

☐ Slag   ☐ FA   ☐ Slag/FA   ☐ RHA/slag   ☐ Waste glass
☐ CTW   ☐ Steel slag   ☐ Ladle slag   ☐ Biomass FA

Fig. 6.3 Relative study number of precursor employed in geopolymers modified with MK evaluated for compressive strength

of MK has a positive effect on the porosity of FA geopolymers as confirmed by around 63.6% of earlier studies. The lower ratios of MK have a benefit in reducing porosity. Contrarily, the introduction of MK has an adverse effect on the porosity of ladle slag geopolymers. The introduction of MK into geopolymers decreases shrinkage (when the main precursors were slag or slag/FA), decreases surface resistivity, increases sulphate resistance, decreases rapid chloride permeability, decreases the expansion of alkali–silica reaction and increases acid resistance.

In actuality, around 37.5%, 12.5%, 10.58%, 6.73%, 6.73%, 4.81%, 3.85% and 3.85% of the references that employed MK as a part of the starting material focused on compressive strength, flowability (workability), porosity, flexural strength, setting time, water absorption, carbonation resistance and fire resistance, respectively as illustrated in Fig. 6.4. Other properties, such as splitting strength (~1.92%), elastic modulus (~1.92%), acid resistance (~1.92%), autogenous shrinkage (~1.92%), drying shrinkage (~0.96%), corrosion resistance (~0.96%), surface resistance (~0.96%), sulphate resistance (~0.96%), rapid chloride permeability (~0.96%) and alkali–silica reaction expansion (~0.96%) have received less attention. To address this limitation, it is advised to increase the number of studies related to these properties.

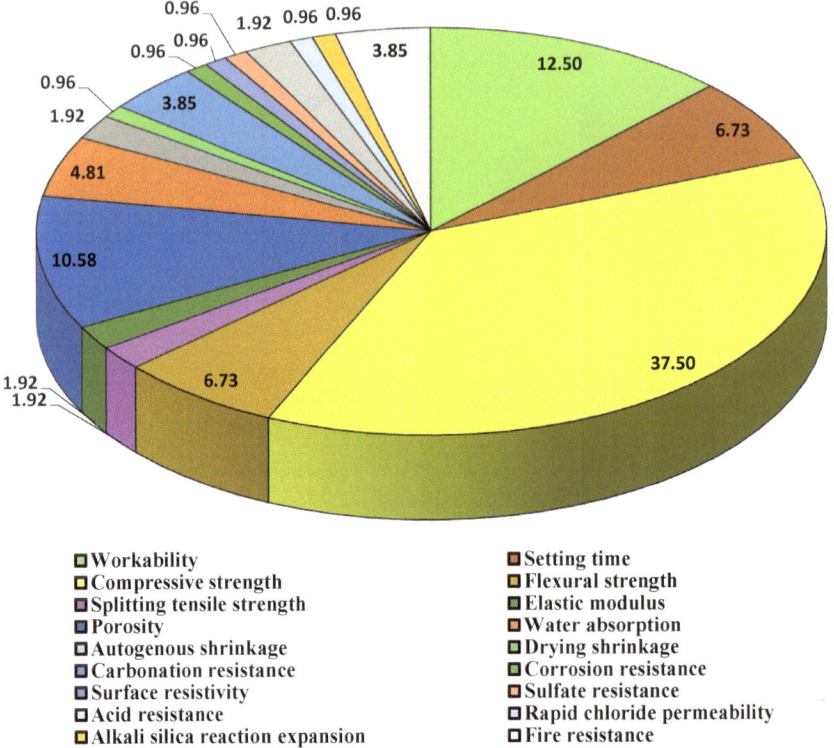

Fig. 6.4 Percentage of studies on the influence of MK on geopolymer properties

References

1. A.M. Rashad, Additives to increase carbonation resistance of slag activated with sodium sulfate. ACI Mater. J. **119**(2) (2022)
2. H. Guo, B. Zhang, L. Deng, P. Yuan, M. Li, Q. Wang, Preparation of high-performance silico-aluminophosphate geopolymers using fly ash and metakaolin as raw materials. Appl. Clay Sci. **204**, 106019 (2021)
3. O. Mahmoodi, H. Siad, M. Lachemi, S. Dadsetan, M. Sahmaran, Development of ceramic tile waste geopolymer binders based on pre-targeted chemical ratios and ambient curing. Constr. Build. Mater. **258**, 120297 (2020)
4. B. El Moustapha, S. Bonnet, A. Khelidj, N. Leklou, D. Froelich, I.A. Babah, C. Charbuillet, A. Khalifa, Compensation of the negative effects of micro-encapsulated phase change materials by incorporating metakaolin in geopolymers based on blast furnace slag. Constr. Build. Mater. **314**, 125556 (2022)
5. V.V. Praveen Kumar, N. Prasad, S. Dey, Influence of metakaolin on strength and durability characteristics of ground granulated blast furnace slag based geopolymer concrete. Struct. Concr. **21**(3), 1040–1050 (2020)
6. M. Gharieb, Y.A. Mosleh, A.M. Rashad, Properties and corrosion behaviour of applicable binary and ternary geopolymer blends. Int. J. Sustain. Eng. **14**(5), 1068–1080 (2021)

7. N. Bheel, P. Awoyera, T. Tafsirojjaman, N.H. Sor, Synergic effect of metakaolin and groundnut shell ash on the behavior of fly ash-based self-compacting geopolymer concrete. Constr. Build. Mater. **311**, 125327 (2021)

8. M.G. Khalil, F. Elgabbas, M.S. El-Feky, H. El-Shafie, Performance of geopolymer mortar cured under ambient temperature. Constr. Build. Mater. **242**, 118090 (2020)

9. O. Burciaga-Díaz, L.Y. Gómez-Zamorano, J.I. Escalante-García, Influence of the long term curing temperature on the hydration of alkaline binders of blast furnace slag-metakaolin. Constr. Build. Mater. **113**, 917–926 (2016)

10. H. Alanazi, J. Hu, Y.-R. Kim, Effect of slag, silica fume, and metakaolin on properties and performance of alkali-activated fly ash cured at ambient temperature. Constr. Build. Mater. **197**, 747–756 (2019)

11. M. Gómez-Casero, C. De Dios-Arana, J. Bueno-Rodríguez, L. Pérez-Villarejo, D. Eliche-Quesada, Physical, mechanical and thermal properties of metakaolin-fly ash geopolymers. Sustain Chem Pharm **26**, 100620 (2022)

12. P. Duan, C. Yan, W. Zhou, Influence of partial replacement of fly ash by metakaolin on mechanical properties and microstructure of fly ash geopolymer paste exposed to sulfate attack. Ceram. Int. **42**(2), 3504–3517 (2016)

13. C. Fu, H. Ye, K. Zhu, D. Fang, J. Zhou, Alkali cation effects on chloride binding of alkali-activated fly ash and metakaolin geopolymers. Cement Concr. Compos. **114**, 103721 (2020)

14. O. Burciaga-Díaz, J.I. Escalante-García, R. Arellano-Aguilar, A. Gorokhovsky, Statistical analysis of strength development as a function of various parameters on activated metakaolin/slag cements. J. Am. Ceram. Soc. **93**(2), 541–547 (2010)

15. A. Buchwald, H. Hilbig, C. Kaps, Alkali-activated metakaolin-slag blends—performance and structure in dependence of their composition. J. Mater. Sci. **42**(9), 3024–3032 (2007)

16. H. Peng, C. Cui, Z. Liu, C. Cai, Y. Liu, Synthesis and reaction mechanism of an alkali-activated metakaolin-slag composite system at room temperature. J. Mater. Civ. Eng. **31**(1), 04018345 (2019)

17. A.B. Pascual, T.M. Tognonvi, A. Tagnit-Hamou, Optimization study of waste glass powder-based alkali activated materials incorporating metakaolin: activation and curing conditions. J. Clean. Prod. **308**, 127435 (2021)

18. H.Y. Zhang, V. Kodur, B. Wu, L. Cao, S.L. Qi, Comparative thermal and mechanical performance of geopolymers derived from metakaolin and fly ash. J. Mater. Civ. Eng. **28**(2), 04015092 (2016)

19. P. Rovnaník, P. Rovnanikova, M. Vyšvařil, S. Grzeszczyk, E. Janowska-Renkas, Rheological properties and microstructure of binary waste red brick powder/metakaolin geopolymer. Constr. Build. Mater. **188**, 924–933 (2018)

20. Z. Chen, H. Ye, Improving sulphuric acid resistance of slag-based binders by magnesium-modified activator and metakaolin substitution. Cement Concr. Compos. **131**, 104605 (2022)

21. O. Burciaga-Díaz, J.I. Escalante-García, Comparative performance of alkali activated slag/metakaolin cement pastes exposed to high temperatures. Cement Concr. Compos. **84**, 157–166 (2017)

22. G.F. Huseien, J. Mirza, M. Ismail, S. Ghoshal, M.A.M. Ariffin, Effect of metakaolin replaced granulated blast furnace slag on fresh and early strength properties of geopolymer mortar. Ain Shams Eng. J. **9**(4), 1557–1566 (2018)

23. A.M. Abbass, R. Firdous, J.N. Yankwa Djobo, D. Stephan, M.A. Elrahman, The role of chemistry and fineness of metakaolin on the fresh properties and heat resistance of blended fly ash-based geopolymer. SN Appl. Sci. **5**(5), 1–17 (2023)

24. A. Sharmin, U.J. Alengaram, M.Z. Jumaat, M.O. Yusuf, S.A. Kabir, I.I. Bashar, Influence of source materials and the role of oxide composition on the performance of ternary blended sustainable geopolymer mortar. Constr. Build. Mater. **144**, 608–623 (2017)

25. E. Furlani, S. Maschio, M. Magnan, E. Aneggi, F. Andreatta, M. Lekka, A. Lanzutti, L. Fedrizzi, Synthesis and characterization of geopolymers containing blends of unprocessed steel slag and metakaolin: the role of slag particle size. Ceram. Int. **44**(5), 5226–5232 (2018)

26. Z. Li, M. Nedeljković, B. Chen, G. Ye, Mitigating the autogenous shrinkage of alkali-activated slag by metakaolin. Cem. Concr. Res. **122**, 30–41 (2019)

27. S.A. Bernal, J.L. Provis, V. Rose, R.M. De Gutierrez, Evolution of binder structure in sodium silicate-activated slag-metakaolin blends. Cement Concr. Compos. **33**(1), 46–54 (2011)

28. S.A. Bernal, Effect of the activator dose on the compressive strength and accelerated carbonation resistance of alkali silicate-activated slag/metakaolin blended materials. Constr. Build. Mater. **98**, 217–226 (2015)

29. Z. Li, X. Liang, Y. Chen, G. Ye, Effect of metakaolin on the autogenous shrinkage of alkali-activated slag-fly ash paste. Constr. Build. Mater. **278**, 122397 (2021)

30. R. Rajamma, J.A. Labrincha, V.M. Ferreira, Alkali activation of biomass fly ash–metakaolin blends. Fuel **98**, 265–271 (2012)

31. M.C. Bignozzi, S. Manzi, I. Lancellotti, E. Kamseu, L. Barbieri, C. Leonelli, Mix-design and characterization of alkali activated materials based on metakaolin and ladle slag. Appl. Clay Sci. **73**, 78–85 (2013)

32. S.A. Bernal, R.M. De Gutiérrez, J.L. Provis, Engineering and durability properties of concretes based on alkali-activated granulated blast furnace slag/metakaolin blends. Constr. Build. Mater. **33**, 99–108 (2012)

33. M.A. Asaad, G.F. Huseien, R.P. Memon, S. Ghoshal, H. Mohammadhosseini, R. Alyousef, Enduring performance of alkali-activated mortars with metakaolin as granulated blast furnace slag replacement. Case Stud. Constr. Mater. **16**, e00845 (2022)

34. S. Bernal, R.M. de Gutiérrez, F. Ruiz, H. Quiñones, J. Provis, High-temperature performance of mortars and concretes based on alkali-activated slag/metakaolin blends. Mater. Constr. **62**(308), 471–488 (2012)

35. S.A. Bernal, R.M. De Gutierrez, J.L. Provis, V. Rose, Effect of silicate modulus and metakaolin incorporation on the carbonation of alkali silicate-activated slags. Cem. Concr. Res. **40**(6), 898–907 (2010)

Chapter 7
General Remarks

1. The introduction of MK into the mixtures may decrease or increase the workability. This mainly depended on the precursor type. Other factors could affect workability such as the SiO_2/Na_2O ratio, activator type and MK amount. When slag was used as the precursor, the introduction of MK increased the workability due to increasing aluminium and silicate content and reducing the number of angular slag particles and calcium content. The same trend was observed when WBP was used as the precursor due to increasing the amount of free water. Contrarily, when FA was used as the precursor, the introduction of MK decreased the workability due to the reduction in the spherical FA particles and the increase in the irregular spiny plate-like MK particles.

2. Earlier studies agreed that the introduction of MK in the mixture decelerated the setting time regardless the precursor type. This is attributed to the reduction in calcium content (if slag was used as the precursor) and the increase in Al_2O_3 and SiO_2 content as well as the slow process of polycondensation with the introduction of MK.

3. The introduction of MK into the mixtures may decrease or increase the density. This mainly depended on the precursor type and MK amount. When slag or RHA/slag was used as the precursor, the introduction of MK increased the density due to the formation of N–A–S–H gel with including SiO_2 and Al_2O_3 from MK in addition to C–S–H gel, which can enhance the microstructure. When FA was used as the precursor, the increased density with the inclusion of MK could be related to the higher density of MK. Contrarily, the introduction of MK into ladle slag geopolymers decreased the density due to the reduction in the specific gravity.

4. There are contradictory findings about the effect of MK on the porosity and water absorption of geopolymers. This typically depends on testing age, activator type/concentration, Si/Al ratio, MK ratio/fineness, precursor type/fineness and curing condition. When slag was used as the main precursor, around 63.6% of the obtained results reported higher porosity when MK was included, whilst

A. M. Rashad, *Metakaolin Effect on Geopolymers' Properties*,
SpringerBriefs in Applied Sciences and Technology,
https://doi.org/10.1007/978-3-031-45151-5_7

only around 36.4% reported lower porosity. To meet this reduction, it is recommended that the ratio of MK not exceed 10%. Higher MK ratios caused the low formation of C–(A)–S–H gel. When FA was used as the main precursor, around 36.4% of the obtained results reported higher porosity when MK was included, whilst around 63.6% reported lower porosity. This depended mainly on the testing age. However, the general trend is lower porosity with a lower ratio of MK. This improvement could be attributed to the MK fineness that can fill the pores. When ladle slag was used as the main precursor, the incorporation of MK showed an adverse effect on porosity. Regarding water absorption, the introduction of 10% and 20% MK into AAS concrete specimens may increase or decrease it. This mainly depended on the testing age and Si/Al ratio. When FA was used as the main precursor, it was better to not increase the ratio of MK by more than 20% to obtain low water absorption.

5. Earlier studies agreed that the introduction of MK in geopolymers decreased autogenous and drying shrinkage. The reduction in the autogenous shrinkage is attributed to the reduction in the self-desiccation of the matrix due to the increased porosity and the reduced chemical shrinkage with the incorporation of MK.

6. Although there are contradictory findings regarding the influence of MK on AAS matrix compressive strength, the majority of these findings (around 45%) indicated a negative effect. Broadly speaking, it is better to not increase the ratio of MK by more than 30% to obtain acceptable compressive strength, but 20% seemed to be better than 30%. This improvement is attributed to the increased formation of N–A–S–H and C–A–S–H gels in addition to C–S–H. Contrarily, higher ratios of MK than 30% may cause an adverse effect on the strength. This is attributed to the incomplete MK reaction in the system due to insufficient activation conditions or the coarser pore structure and lower amounts of reaction product formation with the inclusion of MK.

7. Although there are contradictory findings regarding the influence of MK on FA geopolymer compressive strength, the majority of these findings (around 66.67%) indicated a positive effect. Broadly speaking, it is better to not increase the ratio of MK by more than 30% to obtain acceptable compressive strength. This enhancement is attributed to the fine MK particles that can refine the pore size, increase packing and produce denser microstructure. In addition to the high aluminium ratio in MK. Contrarily, higher ratios of MK may decrease the compressive strength due to the formation of a less compacted microstructure.

8. When MK was incorporated into other precursors, the compressive strength may be decreased or increased. This mainly depended on the precursor type. When waste glass or CWP was used as the precursor, the introduction of MK decreased the compressive strength due to the formation of Si–O–Al bonds that are weaker than Si–O–Si bonds. Contrarily, when steel slag was used as the precursor, the introduction of up to 100% MK increased the compressive strength due to the increase in Al/Na. The same trend was observed when biomass FA or WBP was used as the precursor. When RHA/slag was used as the precursor, the introduction of up to 20% MK increased the compressive strength due to the

fast reaction between silica and alumina compared to the reaction of silica only. When ladle slag was used as the precursor, the introduction of up to 50% MK increased the compressive strength, whilst 75% and 100% MK decreased it.

9. The influence of MK on the flexural strength of geopolymers is mainly affected by the precursor type. When slag was used as the precursor, it was better to not increase the ratio of MK by more than 20% to obtain enhanced flexural strength. This enhancement is attributed to the mitigation of microcracks with the introduction of MK. When FA was used as the precursor, it is better to not increase the ratio of MK by more than 15% to obtain enhanced flexural strength. This enhancement is attributed to the fine MK particles that can refine the pore size, increase packing and produce a denser microstructure. In addition, the fine MK particles can remove the microcracks and produce a more compact microstructure. When WBP was used as the precursor, the flexural strength increased as the MK ratio increased. This enhancement is attributed to the synergistic impact of MK on the slow reaction of WBP. It is better to not increase the ratio of MK by more than 15% in the matrix to obtain enhanced splitting tensile strength. This enhancement is attributed to the fine MK particles that can refine the pore size, increase packing and produce a denser microstructure.

10. Because the available studies are too limited, it is difficult to reach a conclusion about the effect of MK on the elastic modulus of geopolymers.

11. Regarding the influence of MK on the carbonation resistance or corrosion resistance of geopolymers, it is difficult to reach a conclusion due to the limited available studies and the conflicting results.

12. The introduction of 5–10% MK into FA mortars decreased the surface resistivity due to its high aluminium ratio. The introduction of 5–20% MK into FA pastes improved sulphate resistance due to the high Al_2O_3 and SiO_2 content in MK. The introduction of 10% and 20% MK into AAS concretes decreased the rapid chloride permeability. The introduction of up to 70% MK into AAS mortars decreased the expansion of the alkali–silica reaction due to the high aluminium, low calcium and low pore solution alkalinity. The introduction of up to 25% MK increased the acid resistance of AAS matrices. This is attributed to the denser microstructure and the formation of silica-rich gel with including MK. The introduction of MK may decrease or have no effect on the residual strength after exposure to elevated temperatures. It is recommended to increase the number of studies related to this point to reach a clear conclusion.